Surviving the Storm

Autobiographies and Memoirs of Women
from Asia, Africa, the Middle East, and Latin America
Geraldine Forbes, Series Editor

SHUDHA MAZUMDAR
MEMOIRS OF AN INDIAN WOMAN
Edited by Geraldine Forbes

CHEN XUEZHAO
SURVIVING THE STORM
A Memoir
Edited by Jeffrey C. Kinkley

Chen Xuezhao

Surviving the Storm

A Memoir

**Edited with an Introduction by
Jeffrey C. Kinkley**

**Translated by
Ti Hua and Caroline Greene**

An East Gate Book

M. E. Sharpe, Inc.
Armonk, New York
London, England

DISCARDED

An East Gate Book

Copyright © 1990 by M. E. Sharpe, Inc.

Available in the United Kingdom and Europe from M. E. Sharpe, Publishers, 3 Henrietta Street, London WC2E 8LU.

Library of Congress Cataloging-in-Publication Data

Ch'en, Hsüeh-chao, 1906– .
    [Fu ch'en tsa i. English]
    Surviving the storm : a memoir / Chen Xuezhao ; edited with an introduction by Jeffrey C. Kinkley ; translated by Ti Hua and Caroline Greene.
       p.   cm.—(Foremother legacies)
    Translation of: Fu ch'en tsa i.
    Includes index.
    ISBN 0-87332-601-6
    1. Ch'en, Hsüeh-chao, 1906– . 2. Authors, Chinese—20th century––Biography. I. Kinkley, Jeffrey C., 1948– . II. Title. III. Series.
    PL2840.H784Z47713   1991
    895.1′451—dc20
    [B]   90-21300
                                              CIP

Printed in the United States of America

MV   10 9 8 7 6 5 4 3 2 1

Contents

Introduction

Jeffrey C. Kinkley

CHEN Xuezhao was a favorite author in the 1920s and 1930s of young Chinese students like herself. She was not prolific, and she favored short, rhapsodic essays and lyric poems rather than the more popular genre of fiction, but the multiple reprintings of her works (by Lu Xun's presumed antagonist, the Xinyue or Crescent Publishing House, as well as by his friends at Beixin) indicate that she was quite a best-seller for a serious author. Yet, despite her much-acclaimed "antifeudal" views, Chen does not figure prominently in China's standard leftist literary histories. For one thing, she was too shy to get heavily involved in literary organization work and polemics. For all her radicalism, Chen Xuezhao had more of the traditional Chinese woman about her than Ding Ling, whom Chinese males by contrast found masculine. Also, by spending the early 1930s in Paris rather than Shanghai—in literary rather than political pursuits—Chen Xuezhao was in the wrong place at the wrong time to be canonized, despite her friendships with Lu Xun, Mao Dun, Qu Qiubai, and Zheng Zhenduo.

Chen Xuezhao graciously allowed me to interview her when I visited Hangzhou, August 4–6, 1990. Chen Ya'nan and her husband Chen Shumiao provided utmost help in clearing up various points of information, as well as unique copies of precious family photographs for use in this edition. I also thank the Committee on Scholarly Communication with the People's Republic of China and the National Endowment for the Humanities for funding my 1990 trip to China. The footnotes in this book, like the introduction, reflect my judgment and not necessarily the author's.

One senses that, because of her innocently direct personality, they may have taken her to be a "little sister." This suggests another possibility: that she was overlooked because she was a *woman* essayist, in the 1920s, when nonfiction prose was thought to begin and end with Lu Xun and his male cronies (who wrote didactic rather than emotionally charged subjective essays), and in the 1930s, when the familiar essay promoted by Lin Yutang held sway. By the time Chen joined the Communist revolution in Yan'an, the age of great individualist socialist writers was gone forever in China. She did, however, achieve a special elegance of style and dignity of person that served her well when she persevered, as long as she could, in trying to be a writer in the age of Mao Zedong.

Like many a young romantic writer, Chen Xuezhao won her early following in part by creating a lively persona. Looking back on her career now, one can identify as many as four or five personas, fashioned by two completely different authorial personalities: pre-Communist and post-Communist.

When as a girl she joined China's literary ferment following the May Fourth movement, Chen Xuezhao was known as a thoughtful, melancholic young woman alienated from her male-dominated society (a *liulangzhe*, a rover and spiritual exile, as she called herself), but one who still bore dreams of love and liberation, dreams that spoke for the ideals of her generation. As with Yu Dafu and the early Shen Congwen, Chen Xuezhao's wounded soul and plaintive imagination enshrined her as a lonely nonconformist imprisoned in the big city, young and vulnerable but sensitive and humane, pitted against an outer world turned upside down. This evolved into a still more daring and exotic (but also more worldly and intellectualized) persona when Chen Xuezhao actually became an expatriot for eight years, in France, 1927–1935. Sending in essays and stories from abroad, she became the sophisticated "Chinese student over-

seas," possessed of French literary sensibilities and a
heightened moral sensitivity to China's (and the soul's) social ills that was all the more incisive and authoritative
now that she was no longer a sheltered young woman confined by feudal China. Like Ding Ling, she was imagined
by gossip-mongers to be part of a *ménage à trois*, with her
close male friends and intellectual soul-mates Ji Zhiren and
Cai Boling, the Berlin-born son of China's veteran anarchist
revolutionary Cai Yuanpei.*

Chen Xuezhao's third persona, as a daring Chinese incar-
nation of Edgar Snow, was achieved in 1940 almost over-
night, although she had for a decade been publishing news
dispatches in the Tianjin *Dagongbao* (*L'Impartial*) as its special
European correspondent (under pen names, since she was
mostly banned by the Kuomintang [KMT, or Nationalist
Party] even in the 1930s). Like her fellow countryman Fan
Changjiang,† Chen crossed China's great Northwest for a
story; like Snow's, her quest was to see the Communist hearth
at Yan'an, after entering it from the "white areas." Chen became famous all over again, as the author of a rare outsider's
appreciation of the Communist movement and its high morale
in contrast to the Kuomintang's. Jiang Jieshi's (Chiang Kai-
shek's) agents never let her alone after that, so this persona
quite naturally led into her fourth: that of committed Communist writer, remolded at Yan'an by Mao Zedong, his Yan'an
talks, and his rectification campaign—and after 1945, writer
with Communist Party membership, fully purged of her in-

---

*Cai Yuanpei (1868–1940) was chancellor of Peking University dur-
ing the May Fourth movement and became the founder and first president (1928–1935) of the Academia Sinica, the fulcrum of Chinese
scholarship during the Republican era.

†A journalist (1908–1970) who won fame for 1933–1935 reports in the
Dagongbao that were later reprinted in the influential book *The Northwest Corner of China* (1936). Fan headed a branch of the New China
News Agency during the war and went on to assume a position as
veteran Communist journalist and news administration bureaucrat for
the People's Republic.

dividualism and following the party line wherever it might lead.

Not surprisingly, Chen Xuezhao, like Ding Ling, was to be seen writing epic novels about land reform and so forth in the 1950s. This led, again quite naturally, cynics might observe, into a fifth persona—imposed from the outside— as *silenced* Communist writer. Chen Xuezhao was designated a rightist and assigned to employment far beneath her talents. Even so, she kept writing when she could, in hopes of publishing one day. But that was possible only twenty years later. Her memoirs, of which the translation that follows is the second of two volumes, broke that silence.* She began writing them in 1978, responding to the post-Mao thaw a year sooner than most other writers in China, testifying once again to her determination and grit. But best to let Chen Xuezhao tell the story of her life after 1949 in her own words. I shall simply provide background, drawing on the untranslated first volume of her memoirs, *Tianya guike* (A traveler back from the ends of the earth), selections from her early essays, my interviews with the author and her daughter in 1990, and the *Chen Xuezhao yanjiu zhuanji* (Special collected research on Chen Xuezhao), edited by Ding Maoyuan and published in 1983, when it was permissible to celebrate Chen again. By that time she had been rehabilitated and her old friendships with Lu Xun and Mao Dun rediscovered, partly because of the printing of Lu Xun's diaries.†

---

*This second volume was never serialized in a journal. It is entitled *Fuchen zayi* (literally, "Discursive recollections of floating and sinking") (Guangzhou: Huacheng chubanshe, 1981). The "floating and sinking" figure, a metaphor for the ups and downs in Chen's career, repeats the opening words in a calligraphic scroll presented to Chen by Mao Dun in 1980.

†Chen Xuezhao, *Tianya guike* (Hangzhou: Zhejiang renmin chubanshe, 1980, 1982). All but the final chapter was previously serialized in the *Xin wenxue shiliao* (Historical materials on the new literature), 1979–1980. See also Ding Maoyuan, ed., *Chen Xuezhao yanjiu zhuanji* (Hangzhou: Zhejiang renmin chubanshe, 1983).

Chen Xuezhao was born Chen Shuying (also Chen Shuzhang) on April 17, 1906, in the village of Chenjiadai (Chen Family Locks) outside the little Zhejiang county seat on the north side of Hangzhou Bay called Haining. (Haining town lost its county seat status after 1949 and was renamed Yanguan.) Chen's geographic origins were important in regionally conscious China. Zhejiang had long been the center of Chinese learning under the empire. Under the Chinese Republic (1912–1949), not only did the Zhejiang-Shanghai-Jiangsu region produce many of the radical scholars who conspired to overthrow the monarchy, including Cai Yuanpei and Qiu Jin, modern China's archetypal woman revolutionary, whose example Chen never forgot; Zhejiang by itself provided the political, military, and financial leaders that put China back together by force in 1926–28, under Jiang Jieshi.* As one can sense from Chen Xuezhao's memoir, the province was subjected to particularly harsh scrutiny by the new Communist regime for that reason. And yet, such was the cultural power of the lower Yangzi region in the first half of the twentieth century that the new radical generation of Chinese intellectuals and writers—the successors to Cai Yuanpei (himself a non-Communist who often mediated between the radicals and Jiang's Nanjing regime) out to overthrow Jiang Jieshi and establish communism—mostly worked out of Shanghai and were themselves, in embarrassing numbers, mutually supportive "Zhejiang compatriots," from Lu Xun on down. Chen's memoir, like other authentic accounts, allows one to sense how strong particularistic solidarity among fellow regionals—and even among those who had studied in the same foreign country—remained in the New

---

*Mary Backus Rankin, *Elite Activism and Political Transformation in China: Zhejiang Province, 1865–1911* (Stanford: Stanford University Press, 1986); Rankin, *Early Chinese Revolutionaries: Radical Intellectuals in Shanghai and Chekiang, 1902–1911* (Cambridge: Harvard University Press, 1971).

China of Mao Zedong. Lu Xun died in 1936, but Mao Dun moved up to become minister of culture in the People's Republic, while Chen Xuezhao became party branch secretary at Zhejiang University, the flagship institution of higher education in the province. To a point, her memoir provides a thumbnail sketch of Zhejiang intellectual life in the 1950s. Yet she really did not want to take charge of Zhejiang's fractious literati. Having had her thinking "rectified" by Mao Zedong Thought, she wanted to get close to agriculture. In her relatively luxuriant province, that meant tea growing, a trade originally as exotic to her as it is to us.

The family story goes that Chen's grandfather, an economic refugee from Henan, joined the weaving trade in Zhejiang. But he also devoted himself to learned pursuits such as painting and Kunqu opera. Xuezhao's father was a schoolteacher and principal, opposed to the Manchus and old-fashioned practices such as footbinding, and progressive enough to enjoin his surviving sons to see to Xuezhao's education. He died when she was seven; her mother, to whom she was to devote many fond, meditative essays, was confined to bed not long after, so Chen Xuezhao for the rest of her early years was at the mercy of her elder brothers. She was ninth born in the family, without any sisters, and without any younger brothers. Although the elder brothers sent her to school, they also beat her. Xuezhao drew inward. She did odd things like read books at the table while taking meals, and occasionally she responded to force with physical acts of her own. There were many books of classical learning at home, by the ancient poets and historians, so that, despite the brevity of her subsequent formal learning, she did acquire a basis in classical literature. This may have improved the range of her sensibilities and affected her initial choice of genres for her own writing. But Chen, like the best of the other May Fourth writers, would wholly eschew classical figures of speech, preferring to develop China's exciting new creativity in the vernacular.

In the years after May 4, 1919, financial pressure, her own drive, and her desire to be free of her brothers led Chen Xuezhao on a circuit of local girls' schools, first as student and then as Chinese teacher. In 1923 she graduated from the Shanghai Patriotic Girls' School. From that experience she came to know her future close friend Ji Zhiren (son of the principal), the playwright Ouyang Yuqian, and her classmate Zhang Qinqiu, who would marry Shen Zemin. Zhang Qinqiu took Chen Xuezhao to see Zemin's elder brother Shen Yanbing (the famous novelist Mao Dun), and Chen developed a lifelong friendship with Mao Dun's wife, Kong Dezhi (on which, see the memoir).

Chen was already writing amateur compositions. At the end of 1923 she sent in a contribution to a "The New Woman I Hope to Be" essay contest sponsored by the Shanghai *Shibao* (Times). It won second prize. In this, her first printed work, Chen Xuezhao argued for equality between the sexes, insisting that Chinese women, like Ibsen's Nora, must be able to stand on their own two feet economically rather than depend on men. This ideal was to guide Chen Xuezhao for the rest of her life, and not simply from abstract reasoning. In "My Mother" (1925), Chen betrays an obsessive fear of becoming an economic "parasite" within her own unhappy family.

Besides initiating her career as a spokesperson for women's liberation, Chen Xuezhao's first essay led to another longtime friendship, with the *Times* editor Ge Gongzhen, whose nephew Ge Baoquan was to become a Communist statesman noted for promoting cultural relations with the Soviet Union, and likewise a friend of Chen's. She went on to teach at women's normal schools in Anhui and Shaoxing (she was younger than most of her students), in between times living by West Lake in Hangzhou—a quiet spot for writing in those days—or with friends in

Shanghai, or with Zhang Qinqiu and her revolutionary husband. In the summer of 1925 she went to Taiyuan, Shanxi, with Sun Fuyuan's younger brother Fuxi (her sometime boyfriend), to attend one of Tao Xingzhi's Mass Education movement conferences. And she taught and audited classes briefly in Beijing, where Sun Fuyuan introduced her to her fellow provincial Lu Xun. She was already well known to Lu Xun's younger brother Zhou Jianren, an editor at the *Women's Magazine*, to which she contributed regularly.

In 1925 Chen Xuezhao was writing essays and poems on a regular basis and living off her manuscript fees. Little did her brothers know that there was an author in the family, for from the time of her first essay for the Shanghai *Times*, she had adopted the pen name Xuezhao, "Learning from [Liang] Zhaoming [editor of the *Wenxuan* anthology]." *Juanlü* (Wearying travels) (Liangxi, 1925) was her first book, a collection of essays, followed in 1927 by the similarly inspired *Cuncaoxin* (Maternal love, a title alluding to a poem by Meng Jiao) (Xinyue), and *Yanxia banlü* (Travel companion in the twilight mist) (Beixin). When her books went into reprint editions, they financed her passage to France.

Chen Xuezhao's early belles-lettristic writing still bears reading today. In broad form and sentiments, her essays evince many touchstones of classical, domestic sentiment. To quote a few titles: "Maternal Love," "My Mother," "A Rainy Night," "Life at Home," "Forget-me-nots along the Path," "After Weeping," "During Illness," "White Cloud Temple," "The Sounds of Beijing." Some of her essays, which border on the ecstatic, open or close with a poem (in the vernacular, to be sure), while others are diaries, or have a plot, narrated in the first or third person, so that they might be called fiction. "Prose poem," the genre beloved of Zhou Zuoren, is apt for many of her shorter subjective works, so full of imagery and cadence is her language. Chen's traditional themes are developed with modern, intimately subjective and melancholic sentiments, and flavored with such exotic devices as English words appearing in the text in the original Roman alphabet, Christian symbolism, and references to Western philosophers. The authorial personality is that of Chen Xuezhao the free spirit, a dreamy and melancholic young girl beset by the injustices of society and pent up in a psychological and familial prison, struggling alone in the world, but never giving up her sensitivity, her idealism, or her openness to the freshness abounding in the outside world. Thus, to name a few more titles: "The Beginnings of a Life of Wandering," "The Luck of the *Piano*" (the last word is spelled out in Roman letters),* "*Take a Walk*" (a title wholly in English), "Faraway Street Lamps," "Outside the World," "Spring Dreams," "Boredom," and "My Hair" (a familiar essay about hair bobbing). In every case the style is elegant, drawing on imaginative, nonclassical botanical, onomatopoetic, and climatic tropes, set amid short, uncomplicated, typically quite rhythmical sentences in limpid vernacular. There are no classical four-character phrases, no Greco-Roman gods or goddesses, no references to the life of China's new scholar-bureaucrats. *Travel Companion in the Twilight Mist*, with its emphasis on natural scenery rather than the interplay of subjective and social themes, has been nearly repudiated by the author. She wrote it at West Lake, Hangzhou, and claims now that it was unduly influenced by the style of her friend, the younger Sun.

During the Northern Expedition and impending revolution of 1926–27, Chen Xuezhao's life was at a crossroads. She went to visit Mao Dun in Hankou, who offered her editorial work on his newspaper. Other friends asked her to work for the *Dongfang zazhi* (Eastern miscellany), or to do political work for the women's movement. Instead she returned to Shanghai, hoping to go study in France, where

---

*Chen began playing the piano at her old home in Zhejiang, and it remained one of her favorite avocations in Paris.

her close male friend Ji Zhiren had already landed. The white terror in Shanghai in 1927 soon sent nearly all of Chen's leftist friends packing. She might have gone to Moscow, to join Yang Zhihua, the wife of Qu Qiubai, but for the advice of Zheng Zhenduo: "You're a student of literature, not politics; the USSR is good for studying politics." So, with other friends from Sun Fuyuan's Beixin Publishing House (where she had been lodging), she fled in May 1927 to study in France—in part, ironically, to escape pressure from home to marry Sun Fuxi.

With her companions Ji Zhiren and Cai Boling, Chen Xuezhao enjoyed a comparatively idyllic life in Paris. She easily developed her vagabond persona into that of a truly exiled intellectual. She was hemmed in only by the many young Kuomintang partisans in the Chinese community in Paris. They spread rumors about her romantic life and knew enough about her political leanings to keep her from being admitted to a Catholic girls' school outside Paris, on the grounds that she was a Communist. They graduated to posting handbills denouncing her in the Chinese restaurants of Paris after Zhou Jianren's new magazine *Xin nüxing* (New female) printed a letter of hers denouncing their playboy life-styles. But Chen had as her bodyguards the stalwart and scholarly Ji and Cai—one studying music, the other physics—and all this added to the excitement. Chen took French language classes at a division of the Sorbonne and, thanks to an introduction from Ge Gongzhen, was actually able to support herself (for a few months, until her brothers succeeded in getting all payments diverted to them) by sending regular news dispatches to the *Dagongbao*. Since the KMT typically did not allow her to publish under her own name, she not only invented new ones but, she says, altered her style so as to "write like a man." It would prove to be the first major change in her style provoked by external circumstances.

A three-month trip home at the end of 1928, to reclaim

her living wages from her family, fend off their attempts to marry her, and perhaps also demonstrate her independence from Ji Zhiren, who had proposed marriage in Paris, turned out to be a particularly fruitful period for Chen as an author. She lived with Mao Dun and Kong Dezhi, often taking meals with Lu Xun and Zhou Jianren. At this time she penned *Yi Bali* (Memories of Paris) (1929), a collection of descriptive and meditative essays about her home away from home; *Ru meng* (Like a dream) (1929), a book of linked reflective essays whose tone is romantic, like her earlier ones, but also with a few references to the organization of laborers by one "Sister Qin" (as in Zhang Qinqiu); and the novel *Nanfeng de meng* (Dream of the south wind) (1929), about the lives and loves of Chinese students in France. This was, perhaps, her peak, particularly in regard to the belles-lettres that made her famous. From this time forward she would write mostly topical essays and social novels.

Chen Xuezhao's remaining six years in France were personally turbulent. She was still only twenty-three on the trip over, too timid even to leave the boat at ports of call, and too self-conscious to take Ji Zhiren's arm in public, like a Parisienne. But she remained close to Ji and Cai. The three jointly undertook translating work for Lu Xun and Zhou Jianren, notably a version of Turgenev's *Ásya* (translated from the French). Chen's interest evidently ran to Soviet items, for Lu Xun wrote her back that it was all right to send works from capitalist countries, too. The trio further acted as Lu Xun's book agents in Paris, seeking out volumes of art prints for him. And Chen continued to write news reports for the *Dagongbao* until 1931. They fired her when she asked for a raise, so she switched to writing for the Shanghai *Shenbao* "Ziyoutan" (Open forum) and the *Shenghuo zhoukan* (Life weekly). The turbulence, and the need for the raise, came from her hasty marriage shortly before to a boyfriend whom her memoirs refer to only as "H," a Chinese medical student in France. Chen Xuezhao has consented to have him identified, in this English edition, as He Mu, a famous physician specializing in lung disease whose death in June 1990 was front-page news in the Chinese press. One gathers, from the first volume of her memoirs, that Chen initially made friends with him out of pity for his poverty and his tuberculosis. Despite differences in their temperaments, He Mu met her needs for the moment, and she supported him financially. She wrote to Lu Xun, Ge Gongzhen, Zhou Jianren, and Zhang Qinqiu, as if asking for permission and forgiveness; most of her friends responded with shocked silence. Said Zhou Enlai to Ms. Chen in 1942, marriage was the one chapter of her life she had messed up. But marriage may have been her destiny, for she retained a traditional feeling that "a woman should be married," though she evidenced an almost pathological fear of "marrying up." Chen Xuezhao subsequently had a son, but he had to be entrusted to a nursemaid safely away from his tubercular father. The boy died of an illness in 1939, in Chongqing.

Trapped in the south of France—isolated now even from her old Chinese friends in Paris, because of her husband's jealousy—Chen Xuezhao did have one adventure within her misadventure, at Clermont, where there was a university. With the encouragement of Marcel Granet of the Sorbonne, she completed a thesis on classical Chinese *ci* poetry and was awarded a doctorate at the University of Clermont in 1934, under a Professor Carcassonne.* Had she chosen, Chen might then have taken a job at the Sorbonne under Granet.

---

*The thesis was published as Agnès Ho, "Le 'Tse'" (Toulouse: Imprimerie Toulousaine de l'Université de Clermont, n.d.), 54 pp. Ge Gongzhen donated the money to have the thesis printed. Earlier, Chen Xuezhao referred to herself as Agnès Cheng. The language that she spoke with Cai Boling, and still speaks today, is her own Zhejiang dialect. On Granet (1884–1940), see the "Introductory Essay" by Maurice Freedman in Marcel Granet, *The Religion of the Chinese People* (New York: Harper and Row, 1977).

Instead, she returned with her husband to China, in February 1935. There was little to tie her to her old home. Her creative writing was severely diminished now, and her topical writing was virtually unpublishable, due to KMT censorship. Chen Xuezhao found herself frequently tailed by KMT secret agents. But her husband was at least able to make a living for them, after opening a private clinic in Wuxi. The two moved to Nanchang in 1936, where Chen tried to continue writing and translating.

As war broke out and the Japanese pressed on into Central China, Chen, with her husband and baby, made the difficult journey to Chongqing. The trip ended in a surrealistic chance encounter between Chen Xuezhao (who used a false name) and the wife of Jiang Jieshi's intelligence chief Dai Li, in a YWCA restroom. Then the family proceeded to an army hospital in the field. Chen Xuezhao read *Red Star over China*, whose optimistic view of Chinese Communism at Yan'an contrasted sharply with the purposelessness and corruption of their would-be medical unit, which was treating no soldiers. Chen Xuezhao's life was at its pre-1949 nadir. She resolved to join her friend Zhang Qinqiu in Yan'an. With help from the Eighth Route Army liaison office in Chongqing, and invited by the periodical *Guoxun* (National dispatch) to "get a story" on the war effort being waged by the Communists, Chen and her family assumed false identities and slipped into Yan'an in August 1938. Her life was about to change yet again.

Harassed by the Kuomintang and long receptive to the radical ideals of her friends in the Communist Party in China and Moscow, Chen during her ten months in Yan'an felt "as happy and at ease as if I had just come home." She interviewed peasants, intellectuals, and leaders, including Mao Zedong. Zhou Enlai was a neighbor and friend for a time, while he convalesced from an accident. Predisposed to love Yan'an though she may have been, Chen Xuezhao maintained her own individual viewpoint and even continued her French cultural *mission civilisatrice*, by flavoring her piece with, among other things, allusions to Molière and Daudet. But she cleared all her reports in advance with her new mentor, Li Fuchun, deputy head of the party's Central Organization Department, and evidently at her initiative. She seemed a bit disappointed that he only commented "I agree," or "Good," leaving her to ramble on in her own subjective individualism. She had already crossed the line from journalist to committed revolutionary; in her next trip to Yan'an, she would yearn to have her thought rectified, to join the group.

Still, the content and high literary caliber of this imaginative journalism (with a chapter on love and marriage) by a daring young woman who had gone "behind the lines" won her new respect and notoriety, though the KMT killed the serial publication of her book-length *Yan'an fangwen ji* (Interviews in Yan'an) after the first installment. Chen Xuezhao returned to Chongqing in 1939 and was about to see the manuscript published as a book by the Sanlian Publishing House, but it was literally seized off the printing presses. The publisher was arrested, and the original manuscript was confiscated. Fortunately Chen had made a copy. It finally was published in Hong Kong, by Beiji, in 1940, by which time Chen Xuezhao had left KMT territory.

After several uneasy months in the KMT's wartime capital (and after lecturing to a women's group attended, surprisingly, by Mme. Jiang Jieshi), Chen Xuezhao realized that her position in the "white areas" was no longer tenable. A coffee shop opened up across the street from her lodgings, for the sole purpose of spying on her family, she believed. Her suspicions proved well-founded. When she went to Yan'an a second time, she and her companions on the bus were detained and interrogated for nineteen days. Though among those sought by the KMT, she was in disguise and fooled the guards, having torn up her diploma and all other forms of identification with her true name on them, and stuffed them down a rat hole in her lodging. With the help of the bus driver, an underground Communist, her group arrived safely in the Communist capital in the spring of 1940. She wrote an article about her nineteen days of captivity, and some poetry, including an epic piece now lost.

But it was a time for "study," not individual creativity. Officially a member of the Literary and Art Circles' Reserve Committee for Resisting the Enemy (led by Hu Qiaomu), Chen attended Mao's famous May 1942 Talks at the Yan'an Forum on Literature and Art, and quite willingly underwent the great rectification campaign, yearning to rein in her "immature" idealism and to learn to write for a different audience: workers, peasants, and soldiers, as specified by Mao Zedong. She did, of course, become comrades with many major Communist literary intellectuals along the way: Ding Ling, Liu Baiyu, Ai Qing, Ouyang Shan, Xiao Jun, and others. But her main job, after the rectification, was as an editor for page four, the miscellany section (covering science, medicine, agriculture, literature, and art) of the official *Jiefang ribao* (Liberation daily). Ai Siqi* was responsible for page four, under the paper's general editor Lu Dingyi. Other miscellany editors with Chen were Lin Mohan, Bai Lang, and later Zhou Libo. Page four made a political error once, and this made a deep impression on Chen. She also learned to pull in her horns when comrades asked how she could have been so "rich" as to study in France, whether she had gone looking for the Communist Party or it had gone looking for her, and whether or not she hated men. Criticisms of her past individualism must have multiplied because of the divorce she was granted in December 1941. It was a radical thing for a revolutionary

---

*See Joshua Fogel, *Ai Ssu-ch'i's Contribution to the Development of Chinese Marxism* (Cambridge: Council on East Asian Studies, Harvard University, 1987).

woman to do, even in Yan'an (her husband had served with her in the revolutionary base as a doctor), but Chen was able to keep her tiny daughter, Ya'nan. Chen Xuezhao came to worry about the purity of her own class background, and about her many overseas contacts in bourgeois France. Her mentor Li Fuchun (himself trained in France) had warned her to eschew irony; now she learned to be careful in all her speech, after being criticized for an offhand remark to the effect that Mme. Jiang must have observed some things about democracy during her trip to the United States.

Chen Xuezhao devoted herself to reading Marx, Lenin, and Mao. When a production campaign sent the newspaper editors out to do labor, she learned to raise pigs, and to enjoy spinning. She confided to Zhou Enlai that she would like to return to see Europe one day, when the war was over. But even the fall of France was a worry she had to keep from most others, lest they think that she longed to go there. There was talk, from Chen Yi (the future minister of foreign affairs) and others, of using Chen Xuezhao overseas after the war, in international women's league work. She was set to leave for France in 1945, traveling expenses in hand, but a personal conflict with the writer Cao Ming, who was backed by Li Lisan, stopped Chen's departure the day before it was to begin. She would never visit the West again.

In 1944 Chen Xuezhao was assigned to write about model workers, and to review Rice Sprout Song propaganda skits. She heard herself praised in 1945 by Mao Zedong in person, and she was assigned by Chen Yi to write to Cai Boling, to get him to come back to China (he did not). Finally, in July 1945, she applied for party membership. She was immediately accepted, without the usual probationary period—though her friend Deng Yingchao (Mme. Zhou Enlai) advised that she study Lenin on "Infantile Leftism."

After the defeat of Japan, Chen Xuezhao was assigned to work in Manchuria, as editor of page four of the *Dongbei ribao* (Northeast daily). She trekked through the liberated areas of North China and Manchuria en route, which provided material for reportage about common people's work and warfare in her *Manzou jiefangqu* (Journey through the liberated areas) (Shanghai Publishing House, 1949). She also composed a book of short stories about the life of women in New China, and a feminine *Bildungsroman*, the thinly disguised autobiographical novel she called *Gongzuozhe shi meilide* (To be working is beautiful) (Dalian: Xinhua, 1949). The heroine is a literary young woman student who goes to France, marries a Chinese medical student, and later divorces him in China. By this time, Chen Xuezhao's individualistic voice had already changed into a more homogenized, "revolutionary" voice. It was her Francophone friend Zhou Enlai who encouraged her to write it all up, as the saga of one young woman's intellectual odyssey, influenced successively by the May Fourth movement, the Sino-Japanese War, and the revolutionary cause. The heroine shows that revolutionary work is beautiful, as an alternative to home, family, and the joys of youth.

In 1949, Chen Xuezhao, who had for some time been suffering ill health and was anxious to see her home province, petitioned to be allowed to return to work in Zhejiang. In August her wish was granted, and there begins the tale translated below. Responding to "needs" as determined by the party, she as loyal party author wrote a novel about land reform, *Tudi* (The land) (1953), and in 1957 another called *Chun cha* (Spring camellia), about the collectivization of the tea growers. Then she was silenced, condemned as a rightist—yet still she worked on sequels to *To Be Working Is Beautiful* and *Spring Camellia*. Her rehabilitation came in February 1979. She was readmitted to the Communist Party. In recent years she has published two novel sequels, her memoirs, and various works on the literary history she has witnessed. Clearly, this is still the patriotic Communist Chen Xuezhao speaking, as might have been expected in 1978–1980; her list of literary friends at the end of the 1970s reads like a register of those now known as the old hard-liners of 1990: Hu Qiaomu, Kang Zhuo, Lin Mohan, Liu Baiyu, Ma Feng, Ouyang Shan. Her friends in the higher leadership, a virtual Francophone faction, seem mild by comparison: Zhou Enlai, Chen Yi, Nie Rongzhen, Li Fuchun. (An exception is Chen Yun, who remains close to her today.)

The translation of this portion of Chen Xuezhao's memoirs is a project first undertaken by Kai-lai Chung, professor emeritus of mathematics at Stanford University. Professor Chung was moved by Ms. Chen's literary style and drawn into her story by her intimate depictions of the Zhejiang locale, including the scene at Zhejiang University, and particularly the opening reference to the City Station of Hangzhou, which bore many happy associations for him. He initiated the translation by dictation, then had Ti Hua (HUA Ti) and Caroline Greene prepare the bulk of the text, which was revised by myself and Anita O'Brien of M. E. Sharpe. Professor Chung and Chen Xuezhao also kindly supplied proper names of some people and places deliberately left vague in the original Chinese edition. The English text is a slightly condensed version of the original.

The memoir below clearly is written in the spare, relatively guarded voice of Chen Xuezhao in her later years, not the florid lyric voice of the young vagabond of Shanghai and Paris. The reader may well have come to feel that most memoirs originally written in Chinese, including those by famous people maimed by the Cultural Revolution, and even former Red Guards, deal more with events than with inner feelings. (Another peculiarity of Chinese memoirs is their emphasis on urban living space and furniture—but that should come as no surprise to those who have been to China and experienced the extreme scarcity of both even after ten years of reforms.) There is, of course, a great desire on the part of Chinese authors today to "set the facts straight" in the wake of Cultural Revolution distortions, which made so many false allegations precisely about people's private lives and thoughts. On the other hand, Chinese authors' memoir style can hardly be uninfluenced, particularly if they are party members and ex-rightists, by the many "autobiographies" they had to compose during rectification campaigns and background checks. Hence the account-book precision on personnel and participation in party work, even in such matters as loans of money. What is refreshing is that Chen Xuezhao, although naturally intent on setting straight her own story and telling about her famous friends, has not, like some, organized her story so as to mobilize our sympathies around her suffering as a rightist. Status as a rightist was beginning to be seen as a badge of honor when Chen Xuezhao wrote about it, yet she still wondered if she had not in fact spoken out of turn when she was condemned.

As one looks at the literary histories produced in China since 1949, the treatment (or nontreatment) of Chen Xuezhao and her works illustrates that orthodox scholarship has not only dealt unfairly with certain great nonleftist writers, but at times has overlooked even some worthy leftist writers who happened to be outside the designated mainstream and countercurrents defined by literary historians. Western literary critics, on the other hand, will be startled by Chen's own self-revealing displeasure that students at the beginning of the Communist period "were accustomed to studying works line by line and word by word. They were not at all interested in extracting the author's ideology and point of view." Chen Xuezhao's view in 1979 was that much of the harm done even during the Cultural Revolution years was due to failure to follow the correct party line. She emphasized again in 1990 that her memoir in no way criticizes the Communist Party as an organization. One senses her respect for it and her pride in being a member by the way she respectfully refers to fellow party members as "Comrade." It is apparent also that, to a party member, the true locus of decision making is in the hands of party organs. Organs of the government are, by comparison, mere shadows, subordinates of the party. Chen's own predisposition in most things seems to have been to ask permission before taking any step. Truly, she provides a window on a unique moment in the past.

Not that the party was able to call all the shots. One point of interest are the indications in this memoir of how tense and insecure Communist rule was during the early 1950s, at least in Zhejiang. Chen Xuezhao is also a faithful recorder, all the more eloquent because of her detachment, of continuing male selfishness in China. Even among political outcasts, men took the work they pleased, leaving the worst work to women.

To many of us, particularly those who esteem creative writing above all else, Chen Xuezhao's memoir will read like a tale of slow death, at times even self-willed death, of an artist. Yet it is also a tale of strong determination and, above all, self-determination. True to her most fervent ideal, Chen Xuezhao paid her own dues, never allowing men to pay them for her. It was her very anxiousness to escape the traditional female dependency on men that helped propel her into the arms of the collective. She asks not for our pity nor indeed our approval, but only for our understanding.

Surviving the Storm

Chapter 1

IT WAS already nine o'clock in the evening when we got off the train at "City Station" in Hangzhou. Following the instructions of the Shanghai Communist Party Committee, we took two small rooms in the Great Masses Hotel opposite the station. I was fortunate enough to have a small room to myself, but Xu Wenyu, my bodyguard and emissary, had to share his room with other male travelers. We left the few pieces of luggage we had in the hotel and rushed to a small restaurant nearby to eat so that we could return to rest. Everything seemed so different—I felt like a stranger in my hometown!

It was the middle of summer. The heat was stifling even to a southerner like me. Unable to bear it in their tiny room, Xu and his roommates took the wooden boards from their beds to the sidewalk outside to sleep in the open air, right next to the street. Being a woman, I had to suffer in my stuffy little room, where I was attacked by mosquitoes and did not sleep all night. When I arose at dawn, I found Xu sitting out on the stone steps in a daze. Obviously, he had not slept well either.

We went out for breakfast, which was soybean milk and *shaobing*, a flat cake with sesame seeds on top. We then returned to the hotel to retrieve our luggage and pay the bill before hiring a rickshaw to go to West Lake.

Xu refused to ride in it. "It is a means of transportation left over from the old society of exploitation," he declared. Keeping up with the cart was not a problem for him, because it could go no faster than the man who pulled it. I felt uncomfortable in the rickshaw myself, but with my chronically injured right shoulder, I could not have carried a pound of our luggage.

As it was still early when we arrived at the offices of the Zhejiang Provincial Party Committee, we had to wait in the reception room before we could see the secretary of a high-ranking military officer, Commissar Tan Zhenlin. Xu waited outside while I went in to see Commissar Tan. I sat across the table from him, gave him a letter from the central party Organization Department, and repeated what Comrade Liao Zhigao had told me to say. After he had read the letter and asked a few questions, Commissar Tan handed me over to his secretary. I gave the secretary Xu's dossier and asked him to send it to the local party Organization Department. He promptly returned with a letter of introduction and instructions for me to settle temporarily in a party guest house on Lamp Lane, near Bamboo-Stick Lane.

Following the address on the envelope, we found it with considerable difficulty. It was a Western-style storied building. Even though there were few people in it, I shared a room with a female comrade, and Xu roomed with other male comrades. My roommate was a party member who had been on the Long March of 1934–35.

The next day, with permission from the head of the guest house, Xu and I visited my oldest brother's widow. She had raised my second brother's two sons. The younger boy was working in a small silk factory, but the older one was about to lose his job because the corner store where he worked was going out of business. My third brother was dead, too; the only one left was number four, who also was older than I. He was an elementary school principal. His wife was a schoolteacher at a different school. They were family, but I had only seen them once since 1935. During the First National Meeting of Cultural Representatives, I had received a letter from my second elder brother's younger son. Using the address on the envelope, I had found them and shared with them the cloth that was given to me at the meeting. I split it into three shares, one for each family.

A few days passed without any news from the party. I became increasingly anxious as other comrades departed for their tasks, so Xu and I went to the provincial party Organization Department and insisted on going into the countryside. They told me, however, that I must first work with the intellectuals, for the land reform was not yet ready to be put into effect. Shortly after, I attended a meeting of all party members in Zhejiang Province where Commissar Tan gave everyone an emergency assignment—to fight bandits and oppose tyrants. Immediately after the meeting, I reiterated my desire for a job in the country and was sent to a work group in Yiqiao Township in Hang County.

Surrounded by mountains but near a canal, Yiqiao Township is ten or twenty miles from Hangzhou and accessible by two roads and a river. We traveled by boat. Our work group lived in a house that had been abandoned by all its occupants except for an elderly woman when its tyrannical landlord was arrested. Two female comrades and I shared an upstairs room that had a window facing the street. Xu and some other male comrades had a larger room nearby.

In the midsummer heat, mosquitoes beset us even before the sun went down, so we lit repellent incense before going to bed. At midnight, however, the incense burned out, and the insects immediately began their attack. I remembered that a guest in the hostel had suggested that we use netting, so I decided to buy some. But with my limited money supply, I could afford only a small, round-topped netting that was useless because it did not cover my entire body. I did not even bother using it. Xu and I had been away from southern China for so long that we had completely forgotten that summer nights belonged to the mosquitoes. We tried in vain to wrap ourselves in sheets, but the clever creatures still found our necks and faces.

Xu and I always carried guns when we went into the village, because from time to time the bandits showed up. Xu was infuriated that the guns we borrowed from the work group were often inoperative, but I was glad at least to have a blunt object to use in an emergency. When we departed for the South, I had given away my pistol because I had always believed this area to be peaceful. I had forgotten Comrade Liao Zhigao's warning to stay abreast of circumstances. This area had been the cradle for the reactionary Kuomintang for over twenty years.

We knew precious little about the population here and usually chose to visit only the poorest homes. But they were all notably indifferent to us and unwilling to provide any information. They understandably feared the possibility that "the sky might turn"—that there might be a counterrevolution. Having lived under the rule of the Kuomintang for so long, they distrusted Communists. They were quite unlike the peasants in the Communist bases of Yan'an, Shanxi, and Hebei.

I never saw our work group leader, who was in Hangzhou on business. His deputy, a former worker and underground party member, was actually in charge. Our contingent included mostly young intellectuals who had been assigned here following some study at the Zhejiang Cadre School. I felt like a stranger in a foreign land, for I was familiar with neither my comrades in the group nor the local people. The only person I could talk to was Xu. Gradually, however, the group opened up to me, and I began to understand the troublesome situation. They told me the story of a comrade who had recently disappeared after going out alone and unarmed. Several days later, his body was found tied to a pole outside the village. The case had not been solved because the bandits were still rampant and seemingly uncontrollable.

One night, we heard a strange noise that sounded somewhat like a running horse, although it was alternately very strong and very faint. Unable to imagine what it was, we feared that bandits were about to attack us. For safety's sake, I suggested we set up a vigil at the window. At dawn, the group leader and several comrades went outside to investigate and found a broken windmill creaking in the breeze. I realized I had given away my fear and paranoia in this strange place.

A few days later, a deputy county head came to unite the people in rebellion against an unforgiving landlord who had formerly headed the village. The "speaking of the bitter past" meeting, in which the peasants recounted the landlord's injustices to them in his presence, continued for five nights. Previously, when they were asked about their hardships in those meetings, the peasants had named the compulsory military draft as their worst fear. When asked about this involuntary servitude, they avoided inquiry by referring all questions to the head of the village. Now, however, they fearlessly spoke out against the landlord's evils. They had finally stood up for their rights and asked the Communist Party and the people's government to help them gain revenge.

While the struggle against bandits and tyrants was developing, Xu became so ill that he could not even get out of bed. For several days, he removed his jacket and wore shorts, soaking his body in the canal to ease the unbearable heat. He tried to hide this from me, but I found out and attempted to dissuade him from doing it. I was feeling ill, too—at times I shivered from cold, and other times my head swam in scorching heat. Since there was no doctor in the village, we finally decided to request a leave of absence to see one in Hangzhou.

As soon as we arrived there, I sent Xu to the hospital. He had begun to bleed from his gums, nose, and eyes. According to the doctor, he had a serious case of malaria, made worse by the fact that this was his first encounter with the disease. Xu remained in the hospital. The doctor found that I also had malaria and gave me medication. Later, I followed the advice of the provincial party Organization Department and moved into the Xinxin Hotel, a party guest house. After two days, my malaria settled into a consistent pattern: I felt cold and hot every other day, like the "chills" I had suffered as a child. Since the provincial Party Committee's clinic was located near my hotel, the doctor there came to see me and gave me some quinine.

When Xu returned from the hospital a month later, thin and pale, I had already recovered. We asked the Organization Department to allow me to return to the countryside and to let Xu attend the Party School for a while, as his health was not sufficiently restored to bear the southern climate. Because of a need for workers rather than students, the department said that Xu had to work in the Hangzhou railway branch bureau and could study at a more opportune moment. As for my request, the department had to confer with Commissar Tan. He appeared in person to tell me sincerely that sooner or later, when I got used to life in the South, I would have the opportunity to go to the countryside, but that now it was better for me to remain in the city to work with intellectuals in a university. Yet I was still determined to go to the countryside because, without experience, I would not be ready to deal with the intellectuals. With a smile, I insisted that this time I would surely get used to life in the country. It was the first time that I had asserted myself and rejected a party assignment. Comrade Tan urged me to think it over.

Days went by without any news of my assignment, and I became extremely anxious. It was difficult for me to survive a single day without work, so I returned to appeal to the Organization Department. I received the same answer, however: "Commissar Tan would like you to work in Zhejiang University."

During this period, a Soviet-Chinese Association was founded in the province. I recall that Comrade Tan Qilong, first deputy secretary general of the provincial Party Committee, was the chairman of this new organization. I was given a job under its leadership, so I moved to where the association was located. While I persisted in my attempts to return to the country, having little real work to do, I wrote a few articles that were published in the *Zhejiang Daily*, the main provincial newspaper. All the while, Commissar Tan tried to persuade me to work at the university. I understood that he was only concerned with aiding me in my career, and in October, when the school year was about to commence, I finally agreed to take the position. I immediately moved to the campus.

School opened late that year, because the buildings were still being repaired. I used the extra time to find a housekeeper to help me with the shopping and cooking, since one could not bring a bodyguard to a university. Xu had to remain at his railway job.

The former underground party members of the university arranged for my accommodations on campus.* I lived in the front rooms of a small two-family compound with a courtyard. The rear apartment was inhabited by a geography professor who, I heard, was an expert in meteorology. His apartment had three large rooms, while my unit had two smaller ones. My back room served as my living quarters, and I put a little wooden bed against the wall and a small desk and wooden chair by the tall windows, which looked out on the tiny yard. Even though I slept and worked in the same room, I was not cramped, for I did not own much furniture. My room was remarkably tidy and peaceful. I invited my eldest brother's widow to stay in the

---

*Secret Communist Party members who prior to 1949 infiltrated Kuomintang organizations instead of joining up directly with Communist Party and army organizations in Communist base areas.

other room, so that she would no longer have to depend on her two nephews. For her room, I rented some simple furniture from the school—a wooden bed, a square table, and two small benches—and borrowed a little pot for rice from my sister-in-law. My young housekeeper, the daughter of a school employee, worked for us during the day.

Commissar Tan asked me to focus on political classes for the entire school, and specifically on thought reform for intellectuals. My official position was in the Chinese department as professor of third- and fourth-year literary theory and composition. At my request, the party branch allowed me to employ a recent graduate, also a party member, as my teaching assistant. The university now comprised five colleges—Science, Engineering, Agriculture, Medicine, and the Humanities. It had previously contained a law school, too. There were over five thousand members, including faculty, students, and staff.

Before the Liberation in 1949, the reactionary forces of the Kuomintang had shared influence and power over the people with the progressive forces of the Communists. After Hangzhou was liberated by the Communists, however, the Communist Party distributed its dozen underground members among the five colleges to keep tabs on the situation. One was a lecturer, one a teaching assistant, and another was a woman who had just graduated from the history department and worked in the personnel section. All the rest were students. Shortly after I came to the school, the provincial Party Committee decided to bring this party organization in the university out into the open as a party branch and appointed me as the secretary. The deputy secretary was a mechanical engineering student named Xu Zicai, and the organization director was a teaching assistant in the physics department. The overseer of propaganda was a lecturer in the civil engineering department. Some progressives, advocates of communism who were not themselves party members, formed a schoolwide

Political Study Classes Committee. The academic dean, Yan Rengeng, was an expert in economics and served on the committee. He had recently returned from the United States. Belatedly I came to realize that he had been given charge of the school shortly after Liberation. I found him extremely straightforward, honest, and humble. The president of the university, Ma Yinchu, was a well-known economist. He lived in Hangzhou and rarely visited the school. He had been under house arrest by the Kuomintang for many years and was a true supporter of the Communist Party. Neither he nor I had a strong base at the university.

Teaching two courses in the Chinese department kept me quite busy, and I expended even more energy on political lessons for the entire school. I often had to call meetings of the Political Study Classes Committee (later called the Committee on Study), so that we all could exchange opinions and decide what was to be studied. I knew nothing of the political views of the teaching assistant assigned to me in the Chinese department, or his history, and I deemed it improper to question others about him. From the beginning, I did not expect him to help me greatly, but merely to inform me about the school. He was occasionally contemptuous of me, but I did not pay attention, for I felt he was not to blame.

Once, Ma Yinchu was invited to address a full school meeting. The audience was in no mood to listen. Before he could say anything from the dais they began clapping to show their disapproval. They did not stop until I got up and said, "We have already welcomed President Ma, so now we have invited him to speak to us."* I then began to get a sense of the different cliques at the university. I supported diversity among the colleges of the school, but some

*The economist Ma Yinchu was one of pre-1949 China's most eminent scholars. He went on to become president of Beijing University from 1951 to 1960 and became famous for crossing swords with Chairman Mao by crusading to bring population control to China. After Mao's death, Ma's position was officially recognized as the correct one.

egocentric cliques believed that they were elite, superior to everyone else. It was extremely conceited and ignorant of them to think that experts existed only in their particular profession!

One day, I held a meeting of the Committee on Study in a small room connected to the school office. We discussed whether the school should study Chairman Mao's "Talks at the Yan'an Forum on Literature and Art," an article about policy in the intellectual field.* While we were stretching our legs during the intermission, the notoriously arrogant head of the mathematics department† was also in the hallway. He had returned from Japan with his doctorate and was a famous person at the university. When he discovered what we were discussing, he pronounced Mao's speech meaningless and not worth studying.

I could not control my anger at his words, for I knew that if I did not refute this criticism of Chairman Mao, I would feel guilty in front of the party. I sternly told him, "Your view is mistaken. Chairman Mao's article pertains not only to artists and writers but to everyone, because it debates the reform of one's world view. Do you believe that only artists and writers should reform their views while everyone else ignores the problem?"

"We are from the School of Science and will not study that article," he said stiffly.

I replied, "Scientists are not exempted from the reformation of their world view!" The academic dean interrupted and addressed me to avoid a fight. "We shall now continue our meeting." I reentered the room and the head of the mathematics department left in a huff.

When I returned to my apartment after the meeting, I

---

*The locus classicus of Mao Zedong's most important pronouncements on cultural policy. The talks were delivered and transcribed in 1942, during a sweeping rectification movement.

†Su Buqing, who later became president of Fudan University in Shanghai.

thought of my criticism of the mathematician. I had felt that it was my duty to dispute his incorrect opinion, because I believed in open, candid confrontation about discrepancies in others' ideas. I knew I could have been more diplomatic but I lacked the time to soften my words. I knew that both the Science and the Engineering schools had many renowned professors who had studied abroad, like the mathematician, but I did not consider these people uncommon or impressive. According to party policy, however, I was supposed to be kind and polite and to help the intellectuals. Little did I know that when I stood up for Chairman Mao, I was sowing the seeds of future disaster. I thought that knowledge should not be used for personal gain but rather to serve the socialist fatherland and the laboring masses. But the intellectuals did not always agree!

Several days later, a party member secretly told me that the party branch deputy secretary, Xu Zicai, and a few other comrades had written to the provincial Party Committee's Propaganda Department to charge me with lack of respect for elderly professors. They planned to criticize me openly in the press, but their articles never appeared. I was curious as to why my deputy secretary never criticized me to my face. The Party Committee never condemned me or brought me in for questioning. I wondered if it had had second thoughts about my devotion to the party, but I figured that first it would have informed me, and also that my past actions would prove my dedication. History plays no favorites. One day it would prove just what kind of a person Chen Xuezhao was. I was never one to scheme against others or fawn upon the authorities. Under all circumstances, I believed in myself and strove to be an honest and upstanding party member.

One after another, problems of all sorts arose that gave me splitting headaches and kept me extremely busy into the night. After a mere three hours of sleep, I awoke at three o'clock in the morning to prepare for my classes. As early as eight o'clock, people started coming to me with problems. In my class on the modern novel, some students complained that the books written by workers, peasants, and soldiers were simple, monotonous, and without artistic value. When I explained Chairman Mao's literary policy, I could tell that the students did not agree. They eyed each other while refraining from voicing their own opinions. To my surprise, they were accustomed to studying works line by line and word by word. They were not at all interested in extracting the author's ideology and point of view. I wondered if they had ever learned to analyze a literary work. Shortly after, a pistol was found in a public privy near my apartment, and a reactionary slogan appeared on the door to my courtyard during the night. When I left the school in the summer of 1950, I still did not know who had put them there.

The neighbor who lived in my courtyard used to work for a Kuomintang spy organization, the Central Bureau of Investigation and Statistics,* but did not follow the government to Taiwan, because he felt he deserved a more influential position. Often, after a few drinks, he would knock on my door while I was having dinner. He would pound away and denounce the Communist Party leadership until I angrily opened up. Then he would retreat, muttering to himself.

I had heard about many problems in the party branch. Xu Zicai, the deputy secretary, came from a bankrupt landlord family, and as he rarely discussed this with me, I deemed it improper to ask him about it. I had wondered why he could not afford to live at the school, and I noticed that he often had to send money to support his brother and sister-in-law. There were at least two others who could not pay the board at school, so naturally I had to carry the

---

*A notorious "special services" organization under the C.C. clique of the old regime, responsible for intelligence and terrorist operations.

extra load. When I started working at the school, I was no longer compensated with basic living necessities, but was paid a salary. Even though it was adequate, one day I discovered that the rice pot was empty and that I only had a few pennies. My sister-in-law was so angry that she threatened to return to her nephews. When she asked what was for lunch, the young maid solved the problem by buying me five kilos of rice and some vegetables out of her own pocket. She felt sorry for me.

At that time, Xu Zicai and the organization director decided to recruit from the School of Agriculture a new party member in whom they saw great promise. I knew nothing of him but consented because I thought I should agree with the party majority. A female student, C., from the Chinese department also came to me hoping to be recruited. When I first arrived at the school, her father, a professor in the School of Agriculture, wrote me an unusually complimentary letter soliciting my help in her education. Many comrades in the party branch, especially my assistant, thought we should foster the student, but I was skeptical of their adulation. I wanted to be always upright and fair. I did not even acknowledge that Zheng Xiaocang, an education expert who had studied in America, was my cousin, but called him Mr. Zheng as with all the other professors, so no one would think I was partial to him.

On the Chinese New Year in 1950, my comrades in the party branch advised me to pay private ceremonial calls on the department heads and other experts. Having lived for a while in Yan'an, where we arranged group activities to celebrate the New Year, I was not accustomed to the traditions of the old society, but nevertheless agreed because I considered it proper to show my respect for the intellectuals. A female comrade from the Personnel Office accompanied me from family to family to wish them a happy New Year. When we arrived at the home of the dean of the School of Agriculture, his wife informed us that he was not in. He had run upstairs to hide in a neighbor's room to avoid seeing us. We then went up to call on the head of the mathematics department [Su Buqing], whom I had once criticized. His housekeeper also told us that he was out. As we left, we heard people chatting noisily by the upstairs windows, and they started to spit down into the yard with a big "ptooey"! The female comrade was quite annoyed, but I found this rather ridiculous and not worth taking offense.

In the first half of 1950, I went to Shanghai for several meetings about work-study programs and met Comrade Chen Yi.* He encouraged me to speak at one of the major meetings and agreed with my opinion that the work-study program could be tested in the universities since our country's economy had not yet completely recovered from the War of Liberation. My speech was published in the Shanghai *Liberation Daily*.

When summer vacation started, I left the university, because Commissar Tan had vowed that I would return to the country when land reform began. I moved back to the Xinxin Hotel and was again compensated by the basic necessities system.

After I left, Xu Zicai, who was graduating that year, was made a teaching assistant in a major engineering institute in Harbin. A civil engineering instructor was also promoted, because he and Xu had written a derogatory letter about me to the provincial Education Committee. The party always rewarded those it deemed faithful. One day, a comrade from the authorities came to ask me to help Zicai with his travel overseas. I was perplexed as to why he asked me, since I had already left the school, but I had never been a saver, so I gave what I could, about twenty yuan in today's currency. The next evening, he returned to

---

*A top Red Army commander from its inception, Chen Yi became foreign minister of the People's Republic in 1958.

request more money for Xu's brother and sister-in-law. I searched around and found five yuan or so for him, and he did not appear again.

Ma Yinchu had been transferred to become the president of Beijing University before my departure. He had invited the academic dean, Yan Rengeng, to join him because neither man was popular at the school. At the end of the year, Yan came to my hotel to ask my advice, and I urged him to accept the offer. He and his wife moved to Beijing shortly after.

Toward the end of July, after two or three weeks in the guest house, I was ready to leave for the town of Xieqiao in Haining County. I went to say goodbye to the Party Committee's propaganda director. He was ill, but his wife kindly invited me to sit by his bed. After we chatted for a few minutes, I asked for instructions and he told me to go ahead—that my assistant and his girlfriend Miss He, the girl Chinese student C., and the new party member in the School of Agriculture would soon be allowed to follow. The director did not mention, nor did I inquire about, the incriminating letter. Naively, I began to think that the alleged accusation had never been written.

The provincial Organization Department sent an armed guard, Ji Baotang, to accompany me to the country. We were issued a light machine gun, and at the time I carried a pistol of my own, one a leading cadre had once given me. After arriving at the local station, we immediately headed for the district (*qu*) people's government in downtown Xieqiao, the largest town in Haining County after Xiashi and Chang'an. We made our way past the small, crowded shops and presented our letter of introduction to the district director, an agreeable man from the North. He arranged our accommodations at the district government office and persuaded me to remain there temporarily because of the lack of appropriate lodging under the township government. Currently it was putting up its two male comrades in a temple!

The director suggested that we visit nearby Huangdun Village to examine the complex situation there. Following his directions, we crossed the railway to the village, which was spread out along the boundary between three counties. The county lines made the area a good place for bandits to ravage and hide in. Most villages were separated by a mere four miles. Entering the village, we passed between mulberry groves dotted with houses on one side and a deep canal used to water the fields on the other. Like my hometown Yanguan, Huangdun was an industrial crop area growing hemp, wheat, broad beans, and a small amount of rice. The main activity, however, was sericulture, the production of raw silk. In midsummer, the farmers were raising silkworms and the women were picking mulberry leaves in the fields. It was summer again, but this time I got used to the climate and the mosquitoes much quicker than when I had first come, the year before.

After about a month, with the founding of the township people's government, the township director and his copy clerk left the temple and established a dormitory for government personnel in a former landlord's house. Ji Baotang and I also moved there. It was a walled compound, with peach and cypress trees in the courtyard. We had two small rooms across from the kitchen, overlooking a tiny rear courtyard. Once the government moved in, the original landlord's family had to enter and leave by a back door. The township director, from the town of Xucun, was about twenty-four years old and had been wounded and sent home from the East Zhejiang guerrilla forces. The clerk was a young student from the village.

The landlord had passed away long before, leaving his fifty-year-old wife, a concubine in her forties, and two sons who worked in Shanghai. The two women occupied several rooms in a spacious yard. They were generally well-liked, because they had neither used conventional scare tactics to force tenants to pay the rent nor charged high

interest. Some farmers who rented their land had been in-
debted to them for years; the family lived on money that
the sons sent.

Peasants from local villages came to the township
people's government to voice their grievances. They inquired about sons who had been forced to serve in what
was now the enemy Kuomintang Army—were they now in
the People's Liberation Army? They had been afraid to ask
when the army passed through. They asked for help in
getting back daughters who had been obliged to become
concubines for local tyrants. In many instances they complained about the oppressive landlords who had illegally
confiscated their land and mistreated them.

Around the middle of November, my assistant in the
Chinese department and his girlfriend, the girl student C.,
and the new party member in the School of Agriculture
finished their special course on land reform in the provincial capital and joined me in Xieqiao. As soon as they arrived we had an argument, because they wanted me to go
to Jiaxing, site of the province's experimental village for
land reform. Later I discovered that the trip was ordered
by the provincial party Propaganda Department.

During the Taiping Uprising (1851–1864), the inhabitants
of Jiaxing suffered a dreadful massacre, so the majority of
the current population were new emigrants from Shaoxing.
I knew that we should go, but I decided to remain in
Xieqiao because I had my own concerns. I told my new
colleagues that they could do what was convenient for
them. My assistant from the Chinese department went to
Chang'an instead of Jiaxing, but the others stayed with me
in Xieqiao. We lived in the same yard. I had chosen to be
located in my home province not only because I wanted to
participate in the land reform, but also because I was familiar with the various dialects of the region and wanted to
write. It was to avoid being disturbed by family and
friends that I had gone to Xieqiao instead of my hometown,

Yanguan. When I had worked on land reform in Shanxi Province, which was liberated years earlier, I had been unable to write about these enormous changes in the countryside, because I was acquainted with neither the numerous local dialects nor the native customs.

In mid-December, a prefectural Party Committee land reform work team from the county assigned seventeen people to a branch work team in Huangdun Township. Among them were high school and college students, artists, and soldiers. Their leader, a military officer, was about thirty-five years old and had been involved in the Huaihai campaign, the second of three decisive campaigns between the Communist Party and the Kuomintang in late 1948. When the team arrived, they elected a party branch. The leader was chosen as secretary of both the party branch and the Communist Youth League branch. The team selected me to be the organization director and another military comrade to be the propaganda director. Following the elections, we held a meeting to analyze the situation. After a year of work here, the director of the township was extremely knowledgeable. He was born here but had left home at an early age to earn his living working and begging. After being discharged from the army, he returned home but could not find his family. Later the reactionary forces tortured his cousin and uncle and threw them into the sea with many other innocent people, just because they were related to a Communist Army soldier.

According to what we knew, the most powerful and despotic landlord was the former head of the village mutual surveillance (*baojia*) system under the old regime. He lived in a large building opposite the compound now housing the township people's government. He was admonished during the January democratic movement against tyranny, but the peasants and farm workers were still afraid of him. He had earned the bitter hatred of the people by arbitrarily increasing their rents and by sending their sons to serve in the Kuomintang Army. He was also a supporter of the bandit resisters of our new regime.

Since Huangdun was the focal point of the township, I had to be informed of the happenings not only in this village, but in the entire area. To keep on top of the situation, Ji Baotang and I often visited the peasants. A secretary in charge of military activities in the military district government volunteered to come to Huangdun to help us, and he insisted upon sharing a room with Ji, who protested. The head of the township solved the problem by offering to share his room with the new officer, but the latter was not happy with this arrangement.

The new officer followed Ji and me wherever we went. One day, he joined us as we went down to a village, even though the township director had sent him on business elsewhere. We traveled along the railway, which lay between deserted mulberry and wheat fields. At sunset, when we were crossing a short bridge, two men suddenly emerged from a mulberry field and approached us. Ji held his machine gun firmly, prepared to defend me. The officer stared at the men. They slowed down before turning off in a different direction. Naturally, I began to suspect the officer.

Soon after, I learned that he was not a farmer but a vagabond. Following the liberation of Xieqiao, he had led the Communist Party to some charcoal hidden by the reactionaries, so the party rewarded him with this important position. At the time, he owned one of the few pistols around and disregarded the party's discipline. I wondered what he did when he went out alone before breakfast, so one day I sent Ji and the new party member from the agriculture school to follow him. When they came back, they reported that he had met two men under a mulberry tree across the railway and then returned home. One evening he did not reappear, and we heard gunshots behind our house. Not knowing how many of the enemy there were, we nervously prepared to defend ourselves with our measly two guns. At midnight, shooting resumed somewhere near the temple, but no one approached our house. The shots gradually died out around 2:00 A.M.

After discussion, I decided to send Ji to the county people's government to report this event. According to a farmer, the officer and his associates, who acted as spies, had already been arrested. This incident warned us of the danger of devious enemies. For a minimal price they had bought the trust of our leadership!

The struggle had become more intense and complicated. One night, a peasant farmer who had eagerly brought us information committed suicide. When we heard the news, we went to see his family and started an investigation. The man's elderly mother was crying by his body, and his wife was sitting in shock and holding her child's hand. We questioned his neighbors and found that his closest friend was the village militia leader. They were constantly seen talking to each other, but no one knew what was said. That evening, we invited the militia leader to our house. As he entered the room, the work team leader asked him sternly, "Did you cause the death of old Li?"

To our surprise, he admitted, "Yes, I did."

I shot a glance at the team leader and said to the offender, "Let's discuss this calmly." Everyone else left the room. I offered him a bedboard by the window and sat across from him. "You must have had some reason for your action."

His eyes filled with tears. "Please tell me the party will not do anything to me before we clarify the facts." He then began talking and did not cease until it was very late.

I learned a great deal from him, so I said, "You may go home now. Everything will work out." His uncertainty showed in his face.

From what I could gather, there was a complicated story behind his relationship with old Li. The militia leader had a neighbor who had been a sergeant in the reactionary forces. All the villages hated this sly man. He had a son who knew he was near death, so the lad was married to a young woman in hopes of a miracle cure, and in any case to take care of his parents and their home if he died. In the old society, this terrible custom was called "the wedding cure." The young man did die, and his poor widow was still a virgin. The sergeant and his wife, however, enjoyed her presence very much. Since he was aware that the people hated him, the sergeant wanted to recruit the militia leader to protect him by setting him up with his son's widow, but he needed an intermediary. The evil sergeant knew that, for a short time, old Li had worked on the railway for the reactionaries—an act that Li wanted to keep a secret. So the sergeant bribed Li to execute his plan. He also legally adopted his son's widow and promised to remit all the dowry. At first the militia leader opposed marrying the girl, but his mother found her clever and attractive. When the mother saw the girl washing clothes in the river, she invited her to tea so that she could speak with the militia leader. All the while, old Li was gradually convincing him. So the militia leader actually became close to the girl. But he always changed his mind after hearing other people's opinions of the reactionary sergeant at meetings of the people's government.

One day, the militia leader told old Li that he could not marry someone from that family. When Li relayed this information to the sergeant, he threatened to expose to us that Li had once worked for the reactionaries. He must try harder to convince the militia leader. Old Li, who was not familiar with the Communist Party's policy, took this threat seriously. That evening, after telling his wife he was going to the latrine, he hanged himself in a tree. When his wife found his body, it was already too late.

After I had conveyed all this to the work team leader and the township director, we went to question the sergeant's neighbors. At first, fearful of the sergeant's power, they hesitated to tell us anything, but they finally confirmed the militia leader's story. Later, the team leader decided to arrest the sergeant, and I did not protest.*

The sergeant was kept next to the kitchen in a little room without windows. After dinner, the team leader interrogated him, but of course he denied everything, which enraged the leader and all the young comrades. From my room, I heard screaming and crying, and when I saw them beating the sergeant with sticks, I motioned them to stop. As soon as I stopped watching, however, they started to hit him again, and the team leader did not interfere. I was angry because I did not endorse torture, which was illegal. Criminals should be taken to court and sentenced there. I could not sleep until after midnight, when the noise finally subsided. In the morning, it was already very bright when I arose. I walked out of my room and saw the sergeant lying motionless on the floor. The people who were fixing breakfast came out of the kitchen and told me that he was dead. At that moment, the team leader came outside, but I did not utter a word. He went to the director's room to consult with other comrades. When they emerged, they borrowed a board from the landlord's widow. Two male comrades then placed the body on the board and covered it with a sheet before carrying it to the sergeant's house.

A few days later, some of our superiors in the Jiaxing Prefecture land reform team came to investigate. They studied our report and found the work team leader responsible for the sergeant's death. Nevertheless, they decided to delay accusing him because his knowledge of the area was indispensable at that time. Of course, I could not object. Subsequently, a female comrade told me that the team leader was angry at me because the inspection group had criticized him. He had confessed to C., the female student from the Chinese department, "I am polite to Chen Xue-

---

*During land reform, land reform work teams became local governments unto themselves.

zhao only because I respect the party." By that time, I could see right through C. Her specialty was kissing up to the leadership and sowing stories of dissension among the comrades. She had already earned the undivided trust of the team leader. When I heard his opinion of me, I was not upset, because I felt the same way about him even though I would never admit it. We needed to unite for the land reform and to realize that no one was perfect.

The land reform struggle intensified still more. The masses were active. They urged us to remedy their troubles. Also, we had many unfinished tasks, such as measuring fields that were not registered. One clear afternoon, at the request of the peasants, we had two male comrades escort a despotic landlord from village to village so that the people could condemn him. As they were crossing a narrow bridge on the way back to Huangdun, the landlord suddenly jumped into the rushing water and disappeared. One of the men took off his jacket and jumped in after him, but there was no sign of the landlord. Several comrades swam to where the river met the canal but saw nothing. We concluded that the rapid surges of water had swept him into the canal.

In Xieqiao, new conflicts were already taking place. At the end of the year, the War to Resist U.S. Aggression and Aid Korea started. There was a lot of talk among the town leaders about executing the chairman of the old Public Security Committee under the reactionary old regime. They were frightened of what would happen if he returned to power. But we had no concrete evidence; the man was even well liked by the people. In my opinion, it was not imperative that we kill the old man; what dreadful deeds could he do in the rest of this life? Although I could not recall his name, I later remembered that he had taught me Chinese and history in elementary school. I spoke out on his behalf at our meeting even though I could have been accused of being biased. "What do you think of the fact that he has committed no crime and is not despised by the people?" Of course, my words were useless. After our meeting, they held a public trial and dragged him off to be executed. Since he could not move by himself, two soldiers hauled him by his arms to a vacant spot near the railway. His hands and feet were still moving after the first bullet, so they finished him off with a bullet in his head.

A new year came, and the land reform work team was ready to leave. They had distributed all their tools. We had planned to recruit new party members from both the village and our group. The team leader suggested that we enlist C., but his proposal had yet to be approved when he and other comrades from the army were assigned elsewhere. The Jiaxing Prefecture Party Committee sent us a female comrade who had worked in a song-and-dance troupe to be the team leader and the party branch secretary. At this time our main concern was to finish our work.

To my surprise, I received, forwarded from Hangzhou, a letter and a registered parcel containing in two thick volumes a French book from Comrade Bi Jilong, a diplomatic envoy to the United Nations. I had not met him, having only seen pictures of him in the newspaper. In his letter, he explained that the book was a gift from Mr. Cai Boling, and if I had a response, he would take it for me. I had dated Cai Boling when I lived in France, but I had not seen him after I decided to return to China. Without mentioning Cai Boling, I wrote to thank Bi. The work was entitled *The Second Sex* by Simone de Beauvoir and had a card inside the cover that read "À Madame Chen." Writing on a card instead of in a book is a French custom to show respect rather than affection. For two nights, under a little oil lamp, I read the lengthy treatise but could not understand why Cai Boling gave it to me. It argued that marriage does not necessarily imply love, because not all lovers can unite. I presumed that he was trying to console me and that he was married, and that perhaps the French handwriting on the card was his wife's.

When Ji noticed that I had been reading late at night, he mentioned to the township director, whom he thought to be on our side, that I was not resting well. When word got around to the new team leader, she criticized me for devoting my time to French books rather than to the land reform. In 1977, I met her husband on a train to Beijing, and he told me that she despised me for diverting my energy from work to the reading of foreign books at night. As long as I did not get behind in my work, I did not consider night reading such a crime.

The leadership sent Ji and me to a provincial people's government meeting in Hangzhou, but when we returned, the entire atmosphere had changed. C. had been fawning upon the new team leader and denouncing other comrades. The party branch secretary announced a meeting to review my method of distributing the farm tools. That night, I reviewed all my actions and concluded that I had not done anything wrong. I thought, "Let her hold a meeting against me." I had no trouble sleeping that night.

The next afternoon everyone brought his own stool and gathered around a table by the landlord's house. C. was the first to speak. She said that it was an injustice that I had allowed the farmers to use the village's only water cart in turn instead of distributing it among them all equally. Things should be shared, she said. Then the farmers were invited to speak. After a long pause, a man stood up and spoke for them. "Do you think it would be practical to break up the water cart into pieces for distribution?" Patiently, I awaited a response, but the secretary announced the end of the meeting.

Following dinner, C.'s roommate came to my little room, where I was reflecting on the afternoon's activities. She confided that before they had come down to the countryside, the director of propaganda had informed the four comrades who had followed me from the university that even though my knowledge of party policy was greater than theirs, they should struggle against me and eventually persuade me to go to another village. When I had asked the director for instructions before leaving for the countryside, he had been very pleasant and agreeable. However, without any substantial evidence, he was plotting against me because of the letter that C. had helped to write. I wondered why they were secretly scheming against me instead of openly accusing me, because I certainly was able to accept my faults.

While the branch secretary's proposal to take C. into the party was pending, she solicited my approval because I was in charge of recruitment. Everyone knew that I thought she needed further education by the party, so the branch secretary made the decision. I had to agree because I did not want her to think I was prejudiced against C. Therefore, C. became a candidate for party membership along with a man and woman from the village. The man had recruited many young men into the army for the War to Resist the U.S. and Aid Korea, and the woman was the head of the village government.

Under the early spring sun, the farmers busily fertilized their newly distributed land. After a final meeting, the members of the land reform team returned to their previous locations, and Ji and I went back to Hangzhou.

Chapter 2

BACK in Hangzhou, we reported for duty to the provincial party Organization Department and were taken to a hotel on beautiful West Lake. From our balcony we could see the entire lake. Once settled in, I took out my pen to write, but I did not know where to begin. Writing was not as easy as I had thought. My stay in the countryside had been too brief, and I needed time to digest and summarize the experience. It was difficult enough describing the goal of the struggle and party policy, but I had the most trouble analyzing people's actions and emotions.

In the summer, when I had finished a first draft of 200,000 words, I received a letter from my best friends Kong Dezhi, Zhang Qinqiu, and Chen Xuanzhao,* persuading me to go to Beijing to see my daughter Ya'nan. They thought that she was lonely and needed more warmth and attention from me. I departed at the end of the summer and stayed for the first few days at Dezhi's house. She was married to the famous novelist Mao Dun (Shen Yanbing). I shared a third-floor bedroom with Ya'nan, who visited my friends' families on weekends and holidays when I was not in Beijing. One day when my daughter was not around, Qinqiu came over to see me. She and Dezhi told me that once at Sunday brunch, they had teased Ya'nan and told her that she was as skinny as a monkey.

---

*Chen Xuanzhao was an honorary "blood sister" of the author, not a relative. The similarity of names is not from birth.

To their surprise, the child exclaimed, "You are spies! You are tigers for saying that I am like a monkey!" She then ran crying into a corner and refused to eat. My friends had to send her back to her boarding school.

After a few days at Dezhi and Mao Dun's, we moved to Xuanzhao's house on East Biaobei Lane. Xuanzhao's daughter and husband were both extremely fond of Ya'nan. Their daughter went to an agricultural college near Ya'nan's school and often visited Ya'nan and brought her fruit and pastries. Every time Ya'nan stayed at their one-story house, Xuanzhao's husband would mark off her height on a column in the dining room. Ya'nan and I dwelled in three rooms in the inner courtyard. I wrote in a library full of books that was between our bedrooms.

One night I attended a party in Huairen Hall, a meeting place for the top officials at Zhongnanhai in Beijing.* I met Comrade Hu Qiaomu and told him that I was revising a book I had written on the land reform. When he asked about my future plans, I revealed various ideas that I had considered. "Since Zhejiang is famous for its rice and fish, I would like to learn about one or the other. But I always get seasick, and I would not want to cause the fishermen any trouble. I also thought I might want to experience the life of a tea grower, since there are so many in Zhejiang, or . . . "

"Going to the tea fields," he said, interrupting me, "is a wonderful idea!" I was absolutely inspired by his enthusiasm.

Xuanzhao skimmed over my rough draft of *The Land* and commented, "Too much about people, one after another." After I carefully thought about her remark, I decided that she was right—there were too many indistinct characters. I decided to revise the story and shorten it. I then gave a copy to the People's Literature Publishing House. Shortly thereafter, it was published.

---

*Location of China's highest party and government offices; China's Kremlin, and equally inaccessible to the general public.

I spent the Chinese New Year in Beijing and returned to Hangzhou in March. When I reported for duty at the provincial party Organization Department and explained my ideas, the leadership there was supportive and transferred my credentials to the West Lake district Party Committee. I moved temporarily to the Longjing tea plantation and was introduced to the head of the township. There were only two other party members in the entire region: this leader and a man who was being investigated because he had been involved with the enemy. He was later expelled.

The plantation's administrative office was located in a monastery that had built several houses for tourists before the Liberation. I lived in a small room upstairs and ate at the plantation. When I was looking for a woman to help me with the daily chores, Mu Ying, the sister of the deputy township head of the area, volunteered. Since I was out all the time and often could not attend meals, she bought food and cooked it for me on a little charcoal stove in her room across the hall. This poor girl had had a difficult life. As a young child she was sent to another family because her parents could not afford to keep her. This new family did not even like her. Fortunately, Hangzhou was liberated, and after the land reform her parents, with a great deal of trouble, got her back. But since she had not grown up on the tea plantation, she did not know how to pick tea leaves and could only help with housework. Every Saturday afternoon the plantation workers went home for the weekend and left only Mu Ying, myself, and an old monk who lived downstairs in the quiet house.

At this time we were organizing mutual aid teams* in the mountains, which was difficult because the families were so spread out. The deputy township director and I went up and down the mountains every day, and eventually ten teams of five to ten families each were established

---

*A prelude to collectivization.

among the peasants on a voluntary basis. We were the most interested in the team at the bottom of the mountain, which was the smallest. Its leader was a married middle peasant man who had a son in primary school. He was an honest man and always thought of the interest of the country and of his team before making a decision. Without consulting each other, both the director and I referred to this team before any other. Since I was not an official participant in this work, I had a large advantage, because the tea growers would tell me secrets that they would not tell the leaders.

The main argument at that time was whether to stress quality or quantity. Some people advocated growing larger leaves; even though the price per unit would be lower, there would be a greater quantity. Others, including the team leader, were willing to pick the leaves when they were smaller and more tender, so as to sell this lesser quantity for a higher price. Few people understood that the country needed superior quality tea to sell on the international market.

Some of the greedy peasants from the higher mountain teams were only interested in personal gain. The results of the first tea sale, however, proved the worth of the team at the bottom of the mountain. Their net income was by far the highest. I had already become quite familiar with the members of this team, and when they were busy I would help them by tending the fire that toasted the leaves. Even this simple job required skill and training, because it directly affected the quality of the tea. Other times I would help divide the leaves or go up the mountain with the girls to learn how to pick them. The girls had remarkably strict standards as to which ones to select. Even with all the women in the team helping, however, they were short-handed and had to hire experienced pickers from another mountain.

The better I got to know the peasants, the more I loved

and respected them. They believed in their country, revered the party leadership, and had strong self-respect. They had endured the Japanese invasion of Hangzhou without surrendering. When the Japanese were robbing and arresting innocent people and raping the women, the peasants hid in the mountains for days without food.

Summer was near. Mu Ying and I paid a week-long visit to a village across the mountain and stayed in a tea-purchasing center. Even though this village was separated from ours by only a mountain, the situation was notably different. A cadre from the district leadership was in command, and the mutual aid teams were organized by age, into older people's teams and younger people's teams.

Winter was a busy season for the workers at Longjing, for every day they chopped wood for toasting the tea in the spring. Only on rainy or snowy days could they have time to themselves to mend their baskets and clothes. At midnight one Saturday when all the workers had gone home, I was awakened by the glow of a torch outside my window. I instantly went on the alert. Without turning on the light, I silently arose to wake up Mu Ying. We quietly went into the next room, where the security guard lived during the week. He groped for his rifle. Back in my room, we could still see the torch, so we sat and waited in a dark corner. The light did not disappear until dawn, when roosters started to crow and people began chopping wood. Since the light had reflected directly into my window, someone had to have been in the tree. Later, I heard the old monk cough when he opened the door. For some reason, I assumed that his cough was a greeting to someone, but I did not discover what had happened until two years after I left Longjing. The old monk was arrested for numerous incidents of sabotage; once he had tried to burn down the entire temple.

After spending Chinese New Year in Hangzhou, I returned to Longjing. But a few days later, I had to turn back and enter the hospital due to a relapse in my metrorrhagia—hemorrhage of the uterus—probably because I was always rushing about. When I had partially recovered, Comrade Liu Tianxiang, fearing that I had a tumor, urged me to have a follow-up examination in the hospital. The doctor told me to rest for a month, but lying in bed in the hostel, I burned with impatience because it was already time to pick the early tea leaves. After twenty tedious days in bed, Mu Ying and I took a rickshaw to Longjing. The pain in my abdomen was still so acute that I could barely walk.

One day at the beginning of April, as I was writing in my diary, I was told that I had visitors. I went downstairs to the guest room and saw comrades Chen Yi, Nie Rongzhen, and Li Bozhao* sitting at a table, with their security guards nearby. Comrade Chen smiled and said, "Now that *The Land* has been published, may I see a copy?"

"Of course," I answered, "I have one upstairs." When I gave him the book, he read it until lunch time while comrades Nie and Li chatted with me about the Longjing tea plantation. On the way down the mountain, Comrade Chen took a picture of Bozhao and me. At the bottom we got into their car and rode to the military hostel for lunch. I realized that to avoid being greeted by a crowd, they had ridden to Jiuxi and then walked to Longjing, because cars were extremely rare then in that area.

Later that afternoon, Comrade Bozhao came back to see me. She pulled up a chair to my bed and said, "Comrade Chen Yi sent me to invite you to go to Beijing with us to rest, because you are too sick to stay here in Longjing." I smiled and assured her that my health would improve soon. She replied, "Don't try to fool us. You can't even sit up straight, much less walk up and down this mountain.

---

*Li Bozhao, one of the few women on the Long March, in 1930 married Yang Shangkun, whose name is now associated with June 4, 1989.

You should pack, for we leave tomorrow, but now we must have something to eat." I got up, washed quickly, and accompanied her in her car to the military hostel, where we ate dinner with Chen and Nie. Nie gave me a plate of three little fried fish and jested, "Try this marvelous dish. Chen went fishing this afternoon." Chen laughed heartily. During dinner, Nie asked me smiling, "Do you miss France?" I replied, also smiling, "How about you, do you miss it?" "How can I not?" Chen agreed with Nie.

Never when I met with the party leaders, were they false or devious. Instead they seemed unusually humble and sincere, the traits I most admire in a person. They all tried to convince me to go to Beijing with them, but I insisted upon staying. I promised that if I felt weak, I would go to Hangzhou. They finally relented and I returned to Longjing, where I remained until the end of fall.

In the fall of 1950, the Central Research Institute of Literature was founded in Beijing with Comrade Ding Ling as president. Comrade Zhang Tianyi was vice-president but could not take charge of many matters because of illness. The secretary general was Tian Jian, but when he went to Korea in 1951, the deputy secretary general, Kang Zhuo, directed the everyday work. In 1952, Kang Zhuo was succeeded by Comrade Ma Feng.* In the early summer of 1952, the institute gave me the title of professional writer, along with Lei Jia, Bi Ye, and Lu Fei.† I tried to start my next novel, *Spring Camellia*, but I could not condense my many thoughts and felt that I lacked a profound experience to write about. While I was in Beijing, I visited my daughter, who was at the Yuying primary school, one of the few

---

*Ding Ling, Zhang Tianyi, and Tian Jian were among China's most accomplished writers in the pre-1949 era; Ding Ling (1904–1986) is China's most famous Communist woman writer.

†Professional writers draw a salary, whether they write or not. These three, as is not atypical, wrote no important works.

boarding schools in Beijing. I also called on other relatives and friends before returning to Hangzhou in the winter.

After much consideration, I decided to go to Meijia Village, where people were interested in forming cooperatives. When I arrived, I found that my former maid Mu Ying had married, so I employed Zhufeng, the fiancée of a comrade who often drove me to the provincial party meetings. We lived on the top floor of a tea purchasing center where I had such a small room that I could only have a tiny bed, desk, and stool. Zhufeng shared a room with two other girls. As before, we had a small charcoal stove for use when we did not want to eat in the dining hall.

An orphan raised by her grandmother, Zhufeng was hard-working and intelligent. She not only helped me with the daily chores but kept me abreast of affairs in the village. Like Mu Ying, however, her economic situation prevented her from attending school. At Longjing, I had tried to teach Mu lessons in the basic characters. Even after fifteen repetitions of the first lesson, "Workers and Peasants," she still could not remember the words. Since she was not busy in the evenings, I convinced her to go to night school at the plantation. One night, however, she returned from school and cried, "I will never go back there again!" I later found out that someone had mocked her. From then on she would not study at all. I also taught Zhufeng her characters. She studied extremely hard and made unbelievable progress.

With the approval of the leadership of Meijia Village, I was put in charge of the creation of a party branch since there were no other party members in this area. The peasants had various ideas about forming work organizations. The majority were opposed to being grouped by age, preferring to join together voluntarily.

The young people's mutual aid group that was led by L. had more participating families than any other group. Before Liberation, L. and his younger sister, who were orphans, lived by begging, but during the land reform they obtained two rooms and a yard. A gifted speaker who had many friends, L. now became the leader of a cooperative and was being recruited by the party. Many girls had their eye on him, but when he married at the end of the year, no one understood his choice. The girl's father had killed people for the reactionaries and denied his crimes during the land reform. Since some of the victims' families feared the revival of the old rule, he pretended to be well-behaved. Zhufeng gave me this information, which was later confirmed by the people.

After prolonged contact with L., I found that he was only happy when everyone, including the district party leaders, obeyed him. He would refuse to work if they did not fall in line. Once he declined to work because he had run out of rice. He did the same in the summer when his wife gave birth. Both times I bought eggs and rice and sent them to him. I wondered why neither his colleagues nor others who were as poor as he was acted the same way. According to Zhufeng, at the request of his wife he gave all his income to his father-in-law. I did not approve of his recruitment, but the district party branch had already decided to accept him, along with his cousin and another peasant.

After the second tea season, I left Meijia Village to spend a month at Mogan Mountain. Unfortunately, C., who was working for the provincial party Propaganda Department, came with me. I had not seen her since the land reform, but she often sent me documents for the leadership. I had forgotten that her specialty was flattery; she did everything she could to please me, for I had an important position. I could not say no when she asked to accompany me because she had already obtained the leadership's permission.

Summer on the mountain was cool but slightly humid. There was a shortage of water and an absence of vegetables. I could not get started on *Spring Camellia*, so instead I read. The head of the Chinese Writers' Association, Shao Quanlin, and his wife were in the house opposite mine for the summer. I returned to Hangzhou a few days after they left and received a letter inviting me to Beijing.

Whenever I was in Beijing, I visited Premier Zhou Enlai and my "Big Sister Deng," his wife Deng Yingchao. This time he inquired about the tea plantation. I squirmed at his question because I felt that I would be unfaithful to the party if I lied, but I knew my opinion would displease some people. Finally I told him that the tea growers from Longjing often came to ask me how they were to live since the plantation had taken all the fertile land. They were extremely sad because it took seven or eight years for the trees to grow. I then disclosed my opinion to the premier: "The government should control the tea plantations, but it should cultivate new ones in the undeveloped mountains rather than confiscate land from the growers." "You are correct," agreed Premier Zhou. When I later returned to Hangzhou, some of the growers informed me that the plantation had returned their land. They were emphatically grateful to the party.

In September 1954, the leadership asked me to participate in a women's delegation to the Soviet Union to celebrate the anniversary of the October Revolution. Xu Guangping, the widow of the writer Lu Xun, headed the delegation. We traveled across Siberia by train and arrived in Moscow before the festival. The leaders of the Soviet Women's Association welcomed us at our hotel, and we accompanied them to a celebration that afternoon and a ball that evening. We were one of the first delegations to visit Lenin's mausoleum, which had been closed for remodeling since Stalin's death. Stalin's daughter entered ahead of us, and we saw Lenin and Stalin peacefully resting side by side.

Our excursions in Moscow were arranged according to

each person's individual interest. I toured Russian historical landmarks, the tsar's palace, and various cathedrals. Most of my comrades did not understand why a party member would visit cathedrals, but I deemed the structures an integral part of Russian history and did not regret my actions. I also attended the opera, visited farms outside of Moscow, and bought some medicine for my scalp infection at a skin disease center. I was impressed when I found that this new prescription was identical to the one I had obtained from the dermatology department at the Zhejiang Medical School Hospital. Our doctors were quite accomplished and talented.

We then left Moscow to visit Stalin's former residence in Tbilisi. There we climbed up a mountain to get a bird's-eye view of the city. Before flying back to Moscow we visited Sochi, a famous tourist spot on the Black Sea where the weather was still warm. The view of the Black Sea from my airplane was magnificent. It reminded me of the times I had crossed the Indian Ocean and the Mediterranean Sea.

We returned to Beijing at the end of November, and I wrote a few articles about our trip for the *People's Daily*. I also visited Premier Zhou and his wife before going back to Hangzhou. Later I discovered that a female comrade in the delegation had reported my trips to the cathedrals to the party.

Back in Hangzhou, I was ready to write *Spring Camellia* but was bothered by the noise in the hotel. With the help of Cao Xiangqu, an old war buddy of Comrade Shao Quanlin, I found two rooms on either side of a ceremonial hall on the top floor of a convent where two Buddhist nuns lived. They were about forty and twenty-eight years old respectively and were educated, reasonable, and kind. They often borrowed my newspapers to keep up with current events. The older nun always tended the vegetables in her garden, and the younger one sewed and taught at the night school in the village.

Since Zhufeng had gotten married, her husband, Comrade Zhang, presented me with another helper, his neighbor Mo Xiquan. She wrote to express her willingness to work hard, probably because Zhang spoke highly of me in his neighborhood. The nuns were vegetarians, so Xiquan bought provisions at the grocery and cooked them on a little charcoal stove that we had set up downstairs. At first it was very quiet at night, because the nearest houses were on the other side of a small bamboo forest. But one night I heard a sound like stones hitting my window. Xiquan lived in the other room and had to go through a narrow passageway by the ceremonial hall to enter my room. We had never been in the hall because it was closed except when the nuns had to light incense. She was extremely nervous when she had to pass the hall but gradually became more comfortable with it after I told her the savior Buddha was inside.

During that period, Comrade Liu Zhiming, who had been in prison under the reactionary old regime along with a provincial leader, often called on me at the convent. To show his solicitousness, Comrade Shao Quanlin, the head of the Chinese Writers' Association, instructed Cao to buy a small house for me near the convent. Cao found me one for a thousand yuan and offered also to buy a three-room house where soy sauce had once been made. The provincial leadership urged me to consider two other houses instead. The first had three large rooms and was in a good location in the town, but I would have had to pass through the office of a factory to get to it. Also, some of its columns had been eaten by termites. Remote from the city, the other house was near Liutong Temple on Faxiang Lane. It took at least half an hour to walk there from the bus stop. The house was on a hillside, with three rooms in the middle of the hill, and another three facing West Lake to the east at the top of the hill. There was also a small yard with a bathroom and a kitchen. After profusely thanking the

Writers' Association, I immediately moved into this house because I felt I could not refuse the leadership's kind and generous offer. The Writers' Association was able to use its money to buy a house for other vacationers who wanted to write.

My new home had been the summer residence of a capitalist who had gone abroad. It was now owned by the government. I was extremely fortunate to have a deep well in my yard, so I opened the gate twice a day for the people who often had to travel a great distance for water. I chose a room that had a view of both the east and the west. From one side I could see a fruit orchard; from the other, a small road leading to a cliff. I used a second room, which had glass doors, for dining and a third for my study. The only furniture was a long table in the middle room, so with the money I had received for my book I bought a bed, a desk, and some bookshelves from a second-hand store. Comrade Shao convinced me to buy a piano, because he thought I would be lonely living in the country. By chance, I found a used French piano and put it in my study. I seldom played the music I bought, for it revived sad memories. I tried to find some French books in a bookstore and bought the *Sibu Collection*,* but for the most part I read Chinese and foreign classical works and books on the theories of Marx and Lenin.

Shortly after moving in, I was notified that I was to participate in the central group for boosting troop morale attached to the People's Liberation Army, headed by Chen Shutong. Meng Jimao, an orthopedist, and Cheng Yanqiu, a famed Beijing opera singer, were assigned to travel with me. Stopping at each army station to greet the soldiers, we went from Hangzhou to Ningbo. Cheng often performed sections of a Beijing opera. Even though other comrades

---

*A sort of Everyman's Library of classical Chinese learning published in the early twentieth century by the Commercial Press of Shanghai.

tried to hinder him, Chen Shutong was determined to visit the soldiers in Zhoushan because he felt that was what Chairman Mao desired. I was not allowed to accompany him. At the end of this trip, I returned to Meijia Village for a week, staying at the tea purchasing center.

Back in Hangzhou, I found that the other three rooms of my house had been occupied by another woman, a township clerk, because the party deemed it unsafe for either of us to live alone. We contemplated hiring a security guard but decided that was too costly and risky. How could we be certain that he was reliable? I was always nervous at night because there was a cemetery on the other side of the low east wall of the house. All of the coffins had been unearthed, probably by bandits, and bones were strewn everywhere. On the other side of the hill were some farming families, and one of the men whom I knew from the party told me honestly about the dangerous circumstances. I usually could not get to sleep until at least three o'clock in the morning, when I heard the noises of people starting their morning work. Consequently, I began to work late at night and sometimes did not get to bed until dawn.

I made progress on *Spring Camellia* intermittently until summer, when I was summoned to Beijing to study, by the Writers' Association. I left my maid Mo Xiquan in Hangzhou and traveled to the association's residence in the Hepingli district, where I still had two rooms. Comrade Hanzi stayed in the outer room. She, too, was in Beijing for "study." I decided to bring back to Zhejiang my daughter Ya'nan, who had finished primary school, because I did not wish to move to Beijing. My only talent was writing, so I wanted to continue to experience life at the grassroots. I took Ya'nan to live with my youngest elder brother, where his two daughters would keep her company. I left him some money and asked him to urge Ya'nan to take the entrance exam for junior high school. After Ya'nan was settled, I entreated Mo Xiquan to stay with her while I returned to Beijing for two months to study.

Before National Day, I was asked to escort Jean-Paul Sartre and Simone de Beauvoir around the area. An interpreter accompanied us, but the leadership said I could speak to them directly. First we went to the Northeast (Manchuria), where we saw a car factory in Harbin, a steel factory in Anshan, and a coal mine in Fushun. Back in Beijing, we called on another playwright Lao She and Ding Ling, president of the Central Research Institute of Literature. We then traveled to Guangzhou (Canton), where Comrade Ouyang Shan showed us the historical landmarks, introduced us to some fishermen's families, and took us out to eat. Our next journey was to Hangzhou. I invited Ya'nan to my guest house and learned that she had failed the oral part of the entrance exam because she was unfamiliar with the local dialect. When the French visitors came to my house that afternoon for tea, Simone de Beauvoir was extremely surprised to see *The Second Sex* on my shelf. I explained that it was a gift from a childhood friend from France, Cai Boling, and she asserted repeatedly that she would be pleased to deliver a letter for me when she returned to France. Struggling for an answer, I hesitated before I said, "If you see him, try to persuade him to come and see me." Both she and Sartre advised me to visit him in France instead. From what I could tell they deeply loved their country.

Following our stay in Hangzhou, Comrade Zheng Boyong, secretary general of the Provincial Federation of Literary and Art Circles, took us to West Lake. I also took them to Meijia Village. The writers seemed extremely satisfied with the entire excursion. Our last stop was at the Great China Hotel, where Sartre wrote an article that was published in the *Zhejiang Daily*. On the train back to Beijing, I pointed out my hometown. Since they had never seen the silk-making process, they were fascinated by the abundance of mulberry trees whose leaves feed the silk-worms. When they saw the Waibaidu Bridge in Shanghai, they were reminded of a bridge over the Seine. Eventually they returned to the Beijing Hotel, while I stayed in Hepingli.

On the eve of October 1, China's National Day, we went to the national banquet. Sartre sat near Chairman Mao. I introduced both of the French writers to Mao, who was quite warm and cordial to them. Comrade Chen Yi also received them and tried to speak French, but I ended up translating for him. When the writers spoke of Meijia Village, Chen was confused and looked to me for an explanation because it was not a designated or customary stopping place for visitors. I simply told him that I had wanted to take them there. My guests attended the National Day celebration on the visitors' stand at Tiananmen, but I gave my invitation card to another comrade who had longed to go. I was later reproached for this and had to admit carelessness, because Sartre and de Beauvoir had inquired as to my whereabouts. Shortly after, I saw them off at the airport, said goodbye to my friends and relatives in Hepingli, and returned to Hangzhou with Ya'nan.

Although it was an inconvenient location, I set up a bed for Ya'nan in the dining room because I was afraid that if we shared a room I would prevent her from sleeping when I worked late. A few days later, I enrolled her as a boarding student in a private middle school some distance from my house. She came home each Saturday evening and returned to school early Monday morning. Occasionally she took a few hard-boiled eggs back with her, but I was always worried about her health; she didn't eat much. Also, since she was brought up on millet gruel, she was not very strong.

Finally I was able to work on *Spring Camellia*. I emphasized certain characters without relying much on imagination, because I believe a literary work should reflect and never be separated from the reality of its time. At the end of 1955, I decided to finish the first half of my book rather than go to Beijing. I politely refused an invitation from the National Women's Association to tour Italy with Li Dequan* and Guo Lanying. Going to Europe but not to Paris would have disheartened me, so I resolved that someone else should have the chance.

One day Comrade Ding Ning from the Writers' Association called on me. After talking to Mo Xiquan, she was so impressed that she wanted to recruit her into the party. As a result of Mo Xiquan's strong sense of responsibility and dedication to her evening studies, she had improved her political consciousness immensely. Before Liberation, her older sister sold her as a concubine to her brother-in-law so that she would serve them forever. She had two sons and a daughter by him. Immediately after Liberation, her sister dismissed her because of her new belief that no one should have a concubine. Xiquan had nothing to live on, but fortunately she did have a much kinder sister who took her in. When Xiquan came to work for me, this sister kept her two sons, while her mother took care of her little daughter. I gave Xiquan Ya'nan's old clothes to pass on to her children because her sad story touched my heart. I encouraged her to sue her hateful brother-in-law for child support, because of his lucrative financial position. The court decided that he should give Xiquan twenty yuan each month until her daughter reached age eighteen. At first, Xiquan had to go to court each month to collect the money, but later her brother-in-law gave it to her kind sister, who in turn passed it on to her.

I was finally able to finish *Spring Camellia* and send it to the Writers' Publishing House. The Young People's Pub-

---

*Li Dequan, minister of public health, was a women's movement leader and the widow of warlord Feng Yuxiang.

lishing House then asked me to translate two French fairy tales, *The Adventures of Catfish Oscar* and *The Lute and Other Stories*. The first was promptly published, but publication of the second was prohibited when I was denounced in 1957.

That fall, after National Day, Comrade Zhang Yun from the Women's Association brought Mme. Godon from France to Hangzhou and asked me to entertain her. We went to a lecture about women's concerns, in which I was not versed, so I merely sat and listened. We also went to a play. Mme. Godon told me that she was arranging and editing some of her late husband's works. I asked her about the man I used to date, Cai Boling, and she said that he was always busy working but was just fine. She offered to take him a message for me and, as before, I asked her to persuade him to come to China. But to my surprise, she responded in the same way as Sartre and de Beauvoir had: "It would be better for you to go there." The French valued every intellectual asset in their country and never encouraged a talented citizen to leave.

Soon after, one of Cai Boling's college friends from Belgium, Ma Guangchen, came to see me following a meeting in Shanghai. He told me that before Liberation, Boling had been extremely eager to return to China, and that he had received my letter from Yan'an, which Fu Sinian had carried. Cai had decided not to come back, however, because his uncle Zhou Ren wanted him to work for the KMT Academia Sinica. He also could not get in touch with Yan'an. When I heard the story, I could not overcome the heaviness in my heart.

Around that time I wrote two articles, "Jottings in the Countryside" and "Check the Weather," which were published in the *People's Daily* under the pen name Yequ, meaning "Wild Ditch." I had often heard the propaganda director say, "To understand your boss's goals is to be a good worker." I did not agree, however, so I wrote these articles in rebuttal. I opposed the workers' blind submission to their superiors because they were only reinforcing the bureaucracy. I did not advocate insubordination but only that a worker should draw on his or her own analysis and opinion to work better. These articles were later condemned as "poisonous weeds" that undermined the party, its leadership, and the principles of socialism.

Chapter 3

ON FEBRUARY 27, 1957, Chairman Mao gave a speech entitled "On the Correct Handling of Contradictions Among
the People."* Instead of attending the meeting in Beijing at
which the speech was delivered, I asked for a leave of absence. I did, however, read the somewhat different version
that was later published in the newspapers as well as in the
Selected Works of Mao Zedong. I was also absent from the
next meeting, at which Chairman Mao spoke about the
media, but again I read the text of his lecture.

As I was planning to visit the tea-producing area, I received an invitation to return to Huangdun Village. I was
hesitant about the trip until I found out that Shao Quanlin,
party branch secretary of the Writers' Association, and his
wife Comrade Ge Qin were coming to Hangzhou. I decided to remain at home to greet them. One day in late
spring, the couple came up the mountain to see me, and I
accompanied them back to their hostel for a meeting.
Zheng Boyong, the secretary general of the Provincial Federation of Literary and Art Circles, and Cao Xiangqu, a
longtime friend of the couple, were among those present.
The purpose of Shao's trip was to ascertain our opinions of
Chairman Mao's speeches. He visited the president of the
Provincial Federation, Mr. Song Yunbin, at whose house he

---

*A famous text of Maoist theory that may originally have been in-
tended to enliven the Hundred Flowers criticism of the regime, but
which upon its belated public release in June 1957 was put into service
(in revised form) on behalf of a newly launched antirightist movement.

met Huang Yuan. Shao also took a boat trip to West Lake while discussing the situation in our area with Zheng and Cao. I neither was aware of nor participated in these activities.

One day Song Yunbin called a meeting that included the leaders of all the various comrades engaged in propaganda and literary work. I sat facing Shao, in between Ge Qin and Wang Qi, a female journalist from the *People's Daily* whom I had known in Yan'an. Many people approached the lectern to speak, including Cao Canru, literary and art department head of the provincial Party Committee's Propaganda Department, and Lin Chenfu, the deputy secretary general of the Provincial Federation of Literary and Art Circles. Shao and Ge insisted that I speak, but I declined. At the end of the meeting, though, I suddenly wondered if not speaking could be considered to be ignoring the leadership. Consequently I said a few words from my seat. One of my comments was too general: "The provincial Party Committee could have paid more attention to work in literature and the arts." What I meant to say was that opera was excellent in Zhejiang Province, but literature needed more attention. Even though I deemed it improper for me to complain, since I was a writer, I made this remark because Comrade Liu Zhiming had asked me to report his concern. The words just popped out of my mouth.

Two days later, Shao and Ge returned to Shanghai. Then Shao sent a journalist named Rong, whom he had met at the famous writer Ba Jin's house, to interview me. Since Rong had been sent by the Writers' Association secretary, I reiterated my opinion from the meeting. Adding his own interpretation, the journalist later wrote an article entitled "A Chill Follows the Warmth" for the *Wenhuibao* (Encounter news) of Shanghai. Not long after, the official copy of "On the Correct Handling of Contradictions Among the People" was delivered, but I was not allowed to see it because I was already an object of criticism and had been denounced as an unfaithful party member. I do not remember the exact date, but the first accusatory meeting was immediately after the Shanghai journalist's article was published.

The meeting was held in a crowded room on the fifth floor of the provincial government building and chaired by the leading comrade in the provincial Party Committee Propaganda Department. Everyone was asked to condemn me, and some people whom I had never even met vehemently censured me. When a comrade from a modern drama troupe began to praise me instead, he was interrupted and the meeting was adjourned. I was present at this and every other meeting but could not refute the criticism because I had lost my right to speak.

At one meeting the Propaganda Department head declared, "We conducted all of this openly, not secretly." I, however, never knew about any of the meetings until immediately before they were held. He also said that Taiwan had broadcast my comment; only later did I learn that Taiwan had picked it up from monitoring broadcasts by our own New China News Agency. Taiwan radio had said, "Chen Xuezhao, who has always supported the Communist Party, now is having second thoughts." I did not think that this was my fault because I had never been involved with the Kuomintang and my past would confirm my innocence. But at the meetings, C. often lied to incriminate me. Once she went so far as to connect me with the Gao-Rao Group, party members who dissented with Mao and were eventually expelled.* She moved her feet happily under the table while concocting this story, and everyone present seemed to appreciate the information. They thought they had caught a big-time rightist and even an inveterate reactionary.

---

*In 1954, Gao Gang and Rao Shushi were accused of plotting to usurp regional authority and set themselves up in power in Northeast and East China, respectively.

At first, I listened humbly to their denunciation of me because I trusted the party. I had become a member not for personal gain, but to assist the revolution, the country, and the people. Therefore, I believed at first that I deserved the blame and should make amends. But I could not tolerate C.'s blatant fabrications. I stood up and declared, "Go ahead and say whatever you want, for I will never again attend a meeting!" Back at my house, I felt obliged to write Zhou Yang, the chairman of the party's Central Propaganda Department. I also wrote a letter to Deng Yingchao, Premier Zhou's wife, petitioning her to have her husband check to see if I belonged to the Gao-Rao Group. When I was in the Northeast, I had asked a leader why people accepted incriminating evidence about others without proof but would not listen to exculpatory information about them. With an intriguing smile, he had responded, "This is 'class struggle!' " I had been too embarrassed about my lack of comprehension to ask him to explain further, but still I did not understand his words. I resolved to continue studying the works of Marx, Lenin, and Mao.

The next day, some leaders came to chat with me and inspect the books on my shelf, but I was not invited to another meeting for a week. The following week, however, I was summoned to a large meeting at which the leadership changed hands. The old head sat on the side while his successor from Shanghai presided. The speech that I had given at a Zhejiang provincial party representative meeting was found to be in opposition to party principles. I did not refute this assessment, even though it was not I who had written the speech but the secretary and deputy secretary general of the Provincial Literary and Art Federation. I simply agreed with them after I read the document, in which my complaint about the lack of concern for writers was of course incorporated.

Later I saw some wall posters that denounced C. The leadership must have disapproved of her arbitrary declaration that I belonged to the Gao-Rao Group. Evidently this reproof did not worry her, because she still actively criticized me. The newspapers invariably contained articles against me. They could not seem to forget my statement about the party's negligence. The articles also said that I had made a great deal of money from my books, and that I lived in luxury. In truth, I did not have a salary but was treated well by the party because of my title. One article quoted me as having said to C., "Life in the party is complicated." The journalist did not agree. He asserted, "The party is pure and has always been that way." I did think that life in the party was quite complex, but I did not recall that I had expressed my opinion to C. Mo Xiquan and Ya'nan were irate when they read these articles. Xiquan declared that she would no longer cook for visitors, because I had once refrained from eating so that C. would not go hungry.

One day before the Mid-Autumn Festival, both the *Zhejiang Daily* and the *People's Daily* published critical articles about me that seemed to be final verdicts. They classified me as a rightist who opposed the party, its leadership, and socialism. The *People's Daily* also printed a New China News Agency release that named me as a rightist. I received a notice to be present at a party branch meeting. Since the party no longer provided my transportation, I ate an early dinner and went to town to catch the last bus, which left at six o'clock. At the meeting, a female comrade announced my punishment: expulsion from the party and dismissal from all my posts. I would have to live on the income from my books. I promptly rose to leave since I was no longer a party member. The branch secretary and another leader followed me out to a balcony and said, "Your expulsion is not permanent. You may always return if you acknowledge and comprehend your mistakes." My eyes full of tears, I left without responding. I swooned and started walking the wrong way in a daze. It was quite late when I realized where I was, so I begged a rickshaw driver to take me home. When I finally reached my house at midnight, I thanked him profusely and went inside to find Mo Xiquan anxiously waiting for me. I told her that I was going to bed and advised her to do the same.

Lying in bed, I could not fall asleep, but nevertheless kept the light off so that Xiquan would not worry. Over and over I thought, "Now I am useless. I am merely garbage; I only waste the food that the party could give to others." At that point, I did not know whether it was better for me to live or to die. I wished I had gone to Paris in 1937 to work in the Oriental Languages Institute, because the most consequential mistake I could have made was to "escape from political reality." Recalling the two times I had gone to Yan'an to join the party, I knew that I had not erred intentionally. Even though it had expelled me, I resolved to follow the party to the end! Party members were in the minority; it was the masses who made up the majority, and they followed the party. I was pacified when I decided to redeem myself through manual labor.

When I arose the next morning, I wrote a petition to be sent to a village for reform through labor and had Xiquan send it immediately. After three days of waiting apprehensively, I received a reply that stationed me in a cultural center in Shaoxing County.* Because of my high blood pressure, I would not be sent to the countryside. When the township clerk came to see me, I eagerly reported my exciting news, but when he heard, his face took on a strange air. In those days, his attitude toward me was different than before, but I never suspected that he was an informer against me. I eagerly awaited the summons to Shaoxing, but day after day it did not come. Instead, I received a

---

*A county in Zhejiang noted for its wine, its bureaucratic expertise under the old dynasties, and having given birth to Lu Xun and Cai Yuanpei.

notice to move out of my house. A few days later, Wang Yunru wrote me a letter to inform me that her husband Zhou Qiaofeng was on his way to Hangzhou, and that he would like to see me. I immediately wrote back that I would not receive him if he came, so he did not appear and I never again heard from either of them.

Several days before I was classified as a rightist, *Spring Camellia* was published, and I boldly sent copies to my friends and respected colleagues. My friend Xuanzhao wrote to tell me that she had seen it in Beijing bookstores, but when I asked Mo Xiquan to look for my book in Hangzhou, she returned disappointed and empty-handed. Comrade Zhang Xi, the new secretary general of the Chinese Writers' Association, wrote to ask where I wished to go, but I requested only that he send my credentials to Hangzhou. I told him, "I made my mistakes here, so I will rectify them here."

It was extremely urgent that I move out, but I did not know where to find another room. Often people who had spare rooms were hesitant to lease them because they were afraid of being called capitalists. Xiquan and I finally decided to move in temporarily with my sister-in-law, whom I had stayed with when I was required to attend the accusatory meetings. Ya'nan came to visit one or two nights a week, but she was not very happy. The children at school made fun of her, because I was considered a rightist.

When I wrote to Kong Dezhi, she answered right away, begging me to allow her to send me some money. On the envelope, she did not write my name but "Aunt Chen." I decided to ask her never to write to me again, because I did not want to incriminate her family. I said, "If you send me money, I will return it to you at once." I knew that she would deem me unfeeling, but she did not know that I had to start the letter over and over because it was soaked with tears.

I was still searching for a permanent residence when I

was told to inspect a house with three large rooms and a kitchen. It was located on the top of a hill across a deep gully from the mountain where the provincial leaders lived, though unfortunately there was no running water, and I would have had to go to the bottom of the hill each time I wanted it. Since I had been expelled from the party, I did not think that such a large house was appropriate. If anything were to happen, I would not want to take the responsibility. Therefore, I rejected it with thanks.

Coincidentally, a rightist from the *Donghai* magazine was expelled from his home near Shihuqiao and sent to the country. He offered for me to take over his dwelling, but his landlord, who knew that I, too, was a rightist, feared that I could not afford it. For two rooms and half a kitchen, the rent was sixteen yuan a month, not including utilities. He finally agreed to rent them to me when a comrade from the Provincial Federation of Literary and Art Circles acted as my guarantor. It was March 1958 before I settled in and started working at the Shaoxing County cultural center. After I moved in, I went to see my sister-in-law, who was so ill that she could not even speak. A neighbor was taking care of her, but I felt that there was nothing I could do. I left five yuan under her pillow and entrusted her to the neighbor. She died shortly afterward.

When I went to the provincial party Organization Department in Hangzhou to have some paper work done, I was given a new letter of introduction that identified me as a rightist. I then packed a suitcase with a few clothes and a bedding roll, and Ya'nan and Mo Xiquan took me to the bus station. The leaders had told me to take Xiquan with me to Shaoxing, but I felt she had no future working for me. Before I left, I gave both of them some money. Along the road I saw wheat fields that reminded me of my younger days teaching at the Shaoxing County Women's Normal School. Thinking of times past made me melancholy because they seemed at least a century ago. Before Liberation, I had faced many troubles and struggles, but I had always kept the faith and tried my hardest. Now, although I did not give up, I did not see much hope for my old age.

When I arrived in Shaoxing after the two-hour ride, I took a rickshaw to the county party Organization Department to report for duty. I was given a letter to introduce me to the director of the cultural center, who was also the party branch secretary and manager of the New China Bookstore. After reading the letter, he directed me to the center. When I arrived, the only person there was another female comrade. Whispering that she, too, was a rightist, she said that I would have to wait to find out where I would be living until the others returned. Then she showed me a restaurant next door where I could eat, and a place to buy water.

I had to wait until the afternoon for the other comrades, an old man and a woman, to come. They decided to put me in a small room by the stairs and have me clean every room in the center each morning. My tiny room contained a wooden board bed, a small wooden table and stool, and one window facing north. I could barely turn around in front of the bed, but to me a single room was ideal. The female comrade whom I had met first lived in a large room opposite mine. The only thing I could tell about her personality was that she was always extremely strict with me and the other female rightist. I did not know what her profession was, because I only saw her walking around to criticize, insult, and demand things from other people. Like many others, she had been promoted through the political movement rather than by her skill or work. Soon after my arrival, the other female rightist was sent to work in the countryside, and the center was moved to a larger house on Houguan Lane. My new room was upstairs and had a door to the next room, where an artist lived. In addition to cleaning the house, I planned all activities and entertainment for the people. I did this job in a small room downstairs that contained only a desk and a stool, and I ate my meals in a neighborhood factory. At the end of July, I obtained permission to return to Hangzhou to collect my books, which were in damp, termite-infested rooms. Later I discovered that an entire box of books that was under the bed had been ruined.

Although her grades were good, my daughter Ya'nan was not allowed to attend the senior high school because of my status as a rightist. At lunch on my second day back in Hangzhou, Mo Xiquan told me with a smile that she and Ya'nan had something to tell me, and asked if I could guess. When I replied that I had no idea what she was talking about, Xiquan removed a letter from her pocket. It was addressed to Ya'nan from her father H. It read, "I have come to Hangzhou on business. I asked many people before I found your address. I suspect you have been unhappy, but I will always love you. In two days, I will leave for a medical meeting in Paris, but I hope to see you when I return to Beijing, even though your mother prohibits me from getting in touch with you."

I had never told Ya'nan about my disastrous marriage because it had almost killed me, but now I had been exposed and was obliged to explain. Ya'nan wanted to write back to scold him, but I encouraged her to ignore him instead, because his actual intention was to get her to correspond. After making two copies of the letter, I sent one to the provincial party Propaganda Department with an explanation of my divorce and kept one for myself. I sent the original in a registered letter to the central party Propaganda Department and requested that they return it to H. and insist that he not write again. I repeated the words that he had used to describe our divorce: "We made a clean break." Ya'nan added a few words to my letter. At that time, H. was giddy with success and knew that since I was a rightist and no longer belonged to the party, I could never go to Paris again. He often tried to make me sad by ridiculing me, but he only intensified my hatred toward him. I would never be able to forget the past.

One day when my daughter was not with us, Mo Xiquan told me that she had seen Ya'nan take a letter out of her pocket and cry silently while reading it. When she asked Ya'nan what was wrong, the girl showed her the letter, which said that I had already made arrangements to send her to live with H. in Beijing. When Xiquan promised her that I would never do such a thing, Ya'nan agreed to let me see the letter. I then said to my daughter, "I would never send you to his house. He has a new family with many children, and you would not be treated as a relative, but as a maid. In return for saving his life when he was in critical condition and serving and supporting him for twelve years, he turned against me and plotted my death. You will be considered a rightist if you remain here, but you are still my daughter."

After ten days at home I returned to Shaoxing with an enormous mosquito netting. The rickshaw driver kindly carried my heavy suitcase upstairs and hung the netting from the ceiling for me. To show my gratitude for his time and energy, I gave him a thirty-cent tip, and he told everyone downstairs about my generosity. There were more people working in the center when I returned. Cao, whom I had not known in Hangzhou, was the wife of the former deputy director of the provincial cultural bureau. She had transferred to Shaoxing the year before, when her husband had been classified as a rightist. She worked in the morning, went home for lunch, and then returned to work until the late afternoon. Others told me that her husband was a teacher in a high school.

A few days later, the clerk at the center told me to look at the accusatory wall posters downstairs. There was an unusually lengthy one saying that I had not changed in the least and was still extremely rightist. It charged me with trying to buy support from the laboring people since I had tipped the kind rickshaw driver for hanging my mosquito netting. Each day, many names of new rightists were printed in the newspapers. Most of the people, however, had not actually denounced the party but were only suspected of having democratic thoughts. This was because the party had failed to meet its quota for rightists that year.

In late fall, Mo Xiquan wrote to inform me that Ya'nan had a severe case of appendicitis and would be hospitalized for a week for an operation. When she was recovering at home, her friends often visited her and sat talking quietly by her bed. They were also children of rightists and were not allowed to attend senior high school. One day when Ya'nan had fully recuperated, Xiquan returned from the grocery store to find the door locked. When the landlord opened the door, she saw a note on the table. Ya'nan had gone with two girlfriends to work in a feed lot. She later wrote to tell Xiquan that she was raising cows, but she never wrote me, and I did not dare write her for fear that I would put her under suspicion. Since Xiquan now had nothing to do, I encouraged her to work at a small workshop where pages were folded for a publishing house. I wrote a recommendation for her, and she got the job. But I still paid her to look after the house because she did not have much money, having just started the job.

Once, I accompanied the director of the county cultural department to visit model workers in the countryside. When he returned, I remained and lived for a time in a small room in the local people's government building. I helped clean the house each morning and continued visiting workers and attending meetings in the afternoon and evening. Back at the cultural center, I wrote some reports on my trip and gave them to the director, who sent them back after reading them, because they obviously could not be published. I later took these writings to Hangzhou and hid them among my trash. Even though my house was searched many times during the Cultural Revolution, they were never found. Among these articles was one entitled "She Is Ahead of Her Time," about a female laborer.

The cultural center was again relocated, this time to a house next to a farming tool factory. I once more had a small room, located between the bathroom and an empty room. No one wanted to live in the vacant room. It was said that when the former landlord owned the house, his wife had hanged herself because she could not bear the ill treatment of her parents-in-law. I was not scared of ghosts but rather of people who were alive. Since there was a well in our yard, the workers asked us to keep open at all times the gate separating our house from the factory, so that they could fetch water. Consequently, it was much more difficult for me to guard the large house alone at night, as was my job, because I could not lock the gate. Several times at night, I arose to walk around and check the doors, so I never slept longer than two hours at a time. But I was already a stranger to sleep and relaxation, because I was constantly preoccupied with the idea that I was a rightist.

One day the clerk rushed up to me and exclaimed that a box that the artist had left in our former house had disappeared. The artist had not been present when we moved, so the clerk wrote him to inquire about the box. The artist wrote back confirming that he had left a box by the door that connected our two rooms. The clerk then reported the theft to the Public Security Bureau, which sent some men to investigate. Every time we met with the factory workers, they would rudely point at me and say, "Just ask this guard what happened!"

I never replied, but one day the clerk said curtly, "She was not the guard of our former house." I could tell that they were offended, so I reinforced my night vigil. One night, everything was fine when I checked the front door at nine o'clock, but when I returned at half past ten, I found a key in the door. I took it out and put it in my pocket. The only other person who had a key was the clerk, and that night, he had come down to check the door at eight o'clock and had not returned.

By that time, the factory workers had told everyone in Shaoxing about the case of the missing box. When I left the center, the townspeople would cry "saboteur" at me in the street. Since the newspapers had said during the antirightist campaign that I was wealthy because of my books, the people thought that I had stolen the box merely to cause trouble. Often when I was walking down the street, a man would approach me and demand, "Give me three hundred yuan, you saboteur!" I would quickly walk away without replying. The comrades from the Public Security Bureau, however, never suspected me, an irony that made me feel that while the party truly understands people, the masses are easily deceived.

Since the comrades from the Public Security Bureau expected me to report any information exclusively to them, I did not mention the key to the clerk. He did not say a word about it to me. That evening when the police comrades came to inquire about any new developments, I recounted the story and gave them the key. I did not speak of my own suspicions, though. I knew that the factory workers were spreading rumors that both the clerk and I were thieves. It was possible that the clerk had intentionally left his key in the door so that if someone did rob the house, I would be blamed.

Too much tension and too little rest had weakened me. When I put my hands in warm water, it seemed as if there were splinters digging into them. My entire body was so swollen that my eyes always appeared closed. The party branch secretary visited me briefly and decided to send me to the Shaoxing County Hospital, where a doctor gave me a comprehensive examination and recommended that I remain in the hospital. With daily injections, the swelling gradually disappeared. I did not know it at the time, but that was the first sign of hardening of the arteries.

When I was classified as a rightist, the party canceled my medical insurance and reduced my food allotment to a meager twenty *jin* per month.* This never worried me, because I thought that if I really ran out of money, I could always write to the provincial or central Party Committee. When Mo Xiquan came to see me in the hospital, I surmised that she had run out of money and I asked about her financial situation. She told me that several times she had lent money to her sister and that Ya'nan, who did not have a salary, had asked her to pay the rent. The owners of the feed lot assumed that Ya'nan was wealthy since she was the daughter of a rich rightist. Because of this, Ya'nan wanted to be disassociated from me. She insisted that the money come from Xiquan, not from me, so I gave it to Xiquan, who in turn sent it to her.

When I returned to work after two weeks in the hospital, I was moved to a room by the reading area. I continued to clean the house each morning, register newspapers and magazines in the afternoon, and guard the house at night. I did not see the clerk because he was under investigation. A volunteer told me that the clerk, taking advantage of the absence of the center's director, had embezzled several hundred yuan by reducing the number of newspaper subscriptions in the center. He was eventually put in prison for the winter, expelled from the Communist Youth League, and fired from the cultural center. But the people on the streets still shouted "saboteur" at me. I could not understand this, so I wrote a letter to explain the entire theft case to Comrade Zhou Yang in the Central Propaganda Department. At the end, I expressed my true feelings: "I am disappointed with the human race!"

To my surprise, the reply, signed merely by the department instead of by Zhou, arrived promptly. It said, "There is no need to be so distressed, for everything will soon be

---

*One *jin* equals half a kilogram.

cleared up." A few days later, some leaders from the Provincial Federation of Literary and Art Circles came to reassure me. I later discovered that people had circulated rumors about the theft case throughout the province.

At the beginning of the new year, 1959, the Great Leap Forward campaign to boost the economy started. I myself was assigned to proofread popular books sent by the masses reporting rice yields of over ten thousand *jin* per *mu*.* I found that extremely hard to believe, since there had only been one hundred *jin* per *mu* the previous year. My proofreading work soon came to a close; this was followed by a movement for the mass production of iron and steel. All the intellectuals of Shaoxing gathered on an empty plot of land and the men brought the bone coal for the women to strike. So many people were involved in the process that hammers were out of stock in the market. I spent each morning, afternoon, and sometimes evening in a corner hitting chunks of coal before returning alone to the center late at night. Since they all knew I was a rightist, the other women always saved the largest and hardest chunks for me. In this world, there are very few people who will be kind to those less fortunate than they.

Ya'nan once came to spend the night with me, but we did not talk much. She assured me that everything was fine; according to Mo Xiquan, however, Ya'nan's job was extremely taxing. She continually had to run up and down a mountain to carry water to the cows, and she had to get up in the night to feed them. I later found out that she had visited me because a new party branch secretary at the feed lot had told her that it was not appropriate for us to be estranged. After her visit she began to receive a salary like the other workers.

I helped with the coal production movement until the summer, when I was instructed to go with two other com-

---

*One *mu* equals 0.0667 hectares, or 733.5 square yards, hence about one-sixth of an acre.

rades to raise pigs in a village located ten miles from town. One of my colleagues was an old and rather unsuccessful actor, and the other was a stagehand. We rented our house and the pigsty from the local government and bought several large pigs and two small ones from the peasants in the area, but it was difficult to find enough for the animals to eat. We boiled and fermented some chopped vegetables and lily pads, which was the most nutritious mixture for them. I spent all day chopping and cooking the pigs' food, but the others were extremely lazy. The actor's work was limited to a single glance at the sty. Then he spent the rest of the day lying in bed. The younger man merely carried the food that I had been preparing all day to the pigs. Since we only had one room among the three of us, I had to sleep on a makeshift straw mat on the damp and muddy floor while the other two comrades slept on the bed. Not long after, when swine fever was spreading all over the area, the younger man went to town one morning to ask the leadership for instructions. He returned at sunset and announced that the leaders had recommended that we divide up the pigs, go back to the town, and continue raising them at our respective places of employment. After a quick breakfast the following morning, we took the animals back to town by boat. Naturally, when we distributed them, only two small pigs accompanied me to the cultural center.

When I returned, I was reassigned to a small, dark room at the back of the center. Before I moved in, it had been used as a storage room. It was always damp, the floor of the hallway being invariably covered with water. Starting at the crack of dawn, I was busy all day long taking care of the pigs and cleaning the house and sty. Imagining that they cared for me, too, I grew to love my two little pigs and wished they could speak to me. Often, when I went into the sty, they would touch my legs with their muddy noses to show their affection. They always were well-behaved when I lifted them up to lie on a high straw pile each morning while I replaced their wet straw mats with dry ones. At first, I gave them their food in the same bowl, but I later fed them separately so that they would not bump into each other or be crowded while eating. When I gave them two bowls, I could not control my laughter, because they ran from one to the other to make sure that one did not contain better food.

At the end of November, I received a summons to a Provincial People's Political Consultative Conference in Hangzhou. This was still another one of many antirightist meetings, so upon arrival, I was criticized by the chairman. This meeting, however, was unlike the previous accusatory ones. Mr. Zhang Zongxiang, the director of the Zhejiang Library and an old friend of mine, disturbed the meeting by praising me instead of denouncing me. He portrayed me as humble, hard-working, and eager to learn. After he spoke, everyone chatted for a while and then left.

During the meeting, I ran across Dr. Lin Nengwu, who had treated me when I was ill. I described all my new symptoms to him and said, "Please tell me honestly what you think. I can bear the truth!" He told me frankly that he suspected that my high blood pressure had caused a hardening of the arteries, and he encouraged me to make an appointment with Dr. Yu Debao, the most famous ophthalmologist in Hangzhou. His examination confirmed Dr. Lin's suspicion. I had atherosclerosis, but luckily my corneas were not yet bleeding. Explaining that he suffered from a similar problem, the doctor kindly prescribed some medicine and a modified diet. I believed what I had heard about doctors being able to kill people without spilling a drop of blood, but I had complete confidence in Yu Debao and Lin Nengwu, two of the most good-natured and competent doctors I had ever met.

Back in Shaoxing after the meeting, I continued to eat in the nearby factory, but my workload was reduced since the doctor had written a letter requesting it. My little pigs had already been sent to be raised elsewhere. During this period, I met some nice people in Shaoxing. Once when I went to a public shower room, the attendant immediately recognized me and said, "If nothing bad had happened to you, we would have never had the honor of meeting you." She asked me if I was able to buy all the food I needed, but I did not reply because I was ashamed. At that time, there was a shortage of food. She offered to buy chicken, eggs, and fish for me, so with her help, I improved my diet considerably.

One day, the party branch secretary came to tell me, "Make haste to buy canned ham! Bring a lot of money and buy more than you need. Why do you save all your money? Health is all that counts!" When I hesitated because I was working, he exclaimed, "Hurry up! You have my permission." By the time I reached the market, there were only three cans left, so I bought them all. Later, during the Cultural Revolution, the "rebel faction" in Shaoxing questioned me about the branch secretary because he had allegedly become a "capitalist-roader." I realized that if I spoke highly of him, the rebels would think that he had sympathized with a rightist, so I simply said that I did not know much about him since he had rarely visited the cultural center. Part of this was true; I knew nothing of his background.

It was already the end of 1959. Often in the morning, I would boil some water for drinking while I read articles written by the masses. When the sky was clear, I would sit on a bench in the yard after lunch, because my room was too cold.

I heard some rumors about the theft case, but not until after the New Year did I learn the outcome. There was a security guard at the farming tool factory who had been a soldier in the Kuomintang Army before he was captured and made a soldier by the People's Liberation Army. It was later that he joined the party and became a security guard.

One night, he found the box in the artist's room and returned home to get his wife to help him secretly carry it away. They dug a hole under their bed and hid the box there. During the investigation, he had incriminated many people; I, of course, was the easiest person for him to blame the crime upon. Gradually people noticed unusual things in his family. His eight-year-old son started to wear wool suits and leather shoes, and they began using fine gauze mosquito netting, which his wife had made out of the ones in the box. A Communist Youth League member sent to investigate discovered the origin of these novelties. But the guard denied everything when he was confronted. The man then questioned the guard's wife, and she finally broke down and admitted everything. Because of his bad attitude, the guard was expelled from the party and imprisoned for six years. The entire box was worth three hundred yuan—at the most!

Pleased that the case was closed, I carefully composed a letter to inform Zhou Yang, head of the central party Propaganda Department. I also asked to be reassigned to Hangzhou, where I could attend to my health, since I was free to move again. I promptly received a reply referring me to the provincial party Propaganda Department, so I sent the same letter to Jin Tao, its director. Three mornings after I had sent the letter, a comrade from the bureau came to tell me that I needed to do some paperwork before I returned to Hangzhou. I went through those formalities, said goodbye to everyone at the cultural center, quickly packed, and caught a train to Hangzhou in March 1960. Since it was too late to report for duty at the Organization Department, I took a rickshaw and arrived home at four o'clock. Fortunately, Mo Xiquan was there, and she cooked some rice soup for me even though a neighborhood group had established a communal dining hall and encouraged everyone to eat there. Xiquan also borrowed some vegetables from a neighbor and stir-fried them for me.

After an early breakfast the next morning, I reported for duty at the Organization Department but was sent instead to the provincial cultural bureau to register permanently. There I was told to rest at home for two months. I took my medicine every day, but, sad and bored, I was waiting to die. Yet death would not come.

One morning, someone stuck her head inside the front door and asked, "Does Comrade Chen Xuezhao live here?" When I got up to open the door, I saw my "sister" Xuanzhao and her husband. He told me that they had obtained permission to see me from the party's Central United Front Department, but that they were on a business trip and had to leave the next day. They wanted to know all about Ya'nan and were disappointed that they could not see her. Since sugar was scarce in Hangzhou, Xuanzhao brought me some, as well as medicine to fight high blood pressure. They then insisted upon taking me to West Lake for a boat ride, and I felt obliged to obey. When my "brother-in-law" asked why I had turned down the trip to Italy, I replied without thinking, "Since I could not go to Paris, why should I make myself miserable by going to Rome?" We took leave of each other when we returned to shore, and they proceeded to Xuanzhao's brother's house.

My life was still dismal and tiresome. When I finished reading something, I would not understand or remember anything. I reproached myself for maintaining such a depressed outlook and resolved to pull myself together.

Chapter 4

SINCE Mo Xiquan's time was entirely devoted to her work for the publishing house, I found someone in the vicinity to assist me now and then. She came every afternoon to make a fire since my coal supply was too meager to keep the stove lit all day. I always ate rice soup in the morning and bought something from the dining hall to cook for lunch, but every evening my new maid fixed dinner for me and boiled water for the following day. Often, though, she bought canned food instead of cooking because she had good friends who worked in a grocery. It was extremely difficult to obtain any vegetables because even the most inexpensive ones cost at least one yuan for three *jin*. Once in a blue moon, old vegetables would be rationed per capita in each residence. Each time Xiquan went home for a holiday, I would ask her to buy some vegetables for me, because even though they were more expensive where she lived, they were quite fresh. There was also only a limited supply of pork and a total absence of fish. As soon as a store received a shipment of canned goods, there would be a long line, and many who had spent time waiting would leave empty-handed.*

At the end of April, the provincial Propaganda Department summoned me. After my usual breakfast, I went there on May 1 and was taken to a medium-sized room to join the

---

*In the 1980s, it was acknowledged that China had experienced a famine in the wake of the Great Leap Forward.

provincial secretary general of the party, the deputy director of the party Propaganda Department, and a female
comrade, also a rightist, who had held the propaganda position until she praised someone in a foreign country and
differed with the party at a meeting. The men were sitting
together on a sofa and the woman was on a stool. I was
invited to sit on another stool. The secretary general was
instructing the woman to report for duty that very day in
the reference room of the foreign language department at
Zhejiang University, to prepare herself to teach Japanese.
He then turned to me and asked, "What have you been
doing lately?"

"Mending socks," I replied simply.

"Really? Why can't you take walks or do some reading?" Even behind his harsh tone I could detect his genu-
ine concern and kindness. "You will also go to Zhejiang
University, to work in the library for a year while you prepare yourself to teach a French course."

"No!" I almost begged him. "I would prefer to work
solely in the library. I am afraid I cannot take the teaching
job because I have no experience and have not practiced
my French for years." Thinking that, as a rightist, I would
have been a bad influence on the students, I refused him
politely, but with resolution.

It was still early when the female comrade and I left
party headquarters, so we proceeded to the university. We
first arrived at the personnel office but were sent to a different office and finally to the union offices. While we were
waiting in the hall, I said casually to the other woman,
"This is the first time I will have joined a union, because
we did not have one in the Writers' Association. Also, I
didn't think that rightists could belong." But the person in
charge of the official red seals in the union office was not
there, so we were told to return the following day.

I don't know what the woman reported about me, but
when I went to work in the library the next morning, the

Photographs

Chen Xuezhao as a student in 1928, after her first year of residence in France.

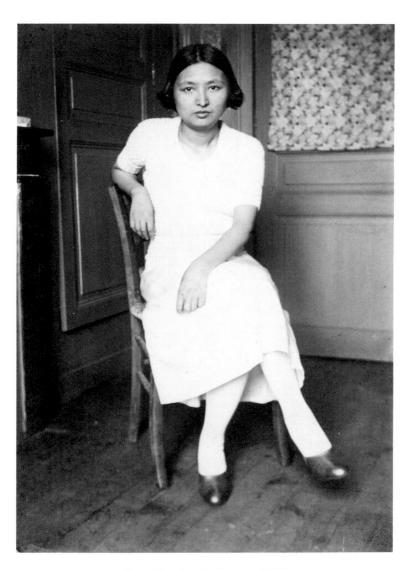

Chen Xuezhao in France, 1930.

Chen Xuezhao in 1947, now a Communist Party newspaper editor and writer in Manchuria.

Chen Xuezhao's old home on the northern outskirts of old Haining town (today Yanguan). Dilapidated at the time of this 1990 photograph, it was abandoned and used for farm storage.

A 1955 touring party at West Lake, Hangzhou. From right to left: Jean-Paul Sartre, Simone de Beauvoir, Chen Xuezhao, and the secretary general of the Zhejiang Federation of Literary and Art Circles.

Chen Xuezhao at a 1980s oral history meeting in Manjuelong, Zhejiang, reminiscing with one of the tea workers who 30 years before had helped inspire her novel *Spring Camellia*.

The author at home, talking to a friend in April 1987.

The author at home in the 1980s, working late at night by lamplight, as is her old custom.

Chen Xuezhao at home in Hangzhou, August 1990.

Right-to-left, front: Chen Xuezhao, Jeffrey Kinkley. Back: The author's daughter Chen Ya'nan and her husband Chen Shumiao. At their home in Hangzhou, a kilometer from West Lake, August 1990.

director's first words to me were: "It is not necessary for you to join the union." I disagreed with him, but, nevertheless, I was silent and obeyed. From then on, I remained at a polite distance from the female comrade. I was always too negligent and casual with my conversation. Often people completely misinterpreted my original intentions. I should have learned my lesson from the antirightist campaign.

Since there was no bus from my house to the university, I had to walk to the library every morning after breakfast. The staff was made up of three groups: acquisitions, circulation, and reference. I edited reference and study materials in the latter group, which contained about ten people. We made registration cards, organized the newspapers and magazines, and were in charge of the reading rooms. My immediate supervisor, not a party member, was also an editor and had specialized in history at Zhejiang Teachers College.

Soon after I arrived, Lin, the vice-president of the university, called me in for a talk. He was a remarkably polite academician. After we briefly discussed my situation, he helped me to regain my entitlement to medical insurance and food rations.

One morning, the leadership sent me to the Hangzhou Hotel to see the writers Liu Baiyu and Sha Ting. I entered a room with two beds. Sha Ting and Ba Jin were waiting for me there. Sha Ting went next door to inform Liu Baiyu of my arrival, and when he appeared, Ba Jin rose from the sofa, greeted me, and left. Baiyu sat next to me on the sofa and Sha Ting sat on a bed. Each of us knew that there was a lot that should be said, but no one knew where to start, so they first inquired about my health. After we talked awhile, I prepared to leave. Sha Ting walked me to the front gate and said, "Take it easy and look at the bright side of things. Take care of yourself!" I was on the verge of tears when we parted.

In July 1960, I received a notice to participate in the Third Cultural Representatives Meeting, which was to take place from July 22 to August 13. In hot pursuit of this announcement, however, was a letter summoning me to the provincial cultural bureau. When I appeared, a vice-director of the bureau instructed me not to attend the meeting, so I did not.

One day late in the spring of 1961, the leadership informed me that Zhou Yang had recently arrived in Hangzhou and had invited me to visit him. The next morning I called on him and his wife, Comrade Su Lingyang in a room on the second floor of the Hangzhou Hotel. They offered me a seat and seemed to have a lot to say. In a slightly reproachful tone, Zhou Yang asked me why I had not written a single letter to him nor to the premier's wife, Deng Yingchao. When he mentioned her, I was no longer able to control myself. Covering my mouth with a handkerchief to tone down my sobs, I burst into tears. I did not tell him that I had not written to them because I felt unworthy of their care and attention. After I was declared a rightist, a leader had told me that I could write to whomever I wished and did not have to show my letters to the party, but that I should never lend money to anyone without permission. I did not write to leaders in the central party apparatus, because I was ashamed, nor to my friends, for fear that I would incriminate them. I remained at the hotel, talking to Zhou Yang and Su Lingyang until after lunch.

A few days later, the school administration sent me to the provincial Propaganda Department, where a deputy director greeted me politely and informed me that my salary would resume the next month. I thanked him profusely but said, "I still have some money left from the profits from my writing, so I can wait for my salary until my savings run out."

He replied, "No. You should be paid for your work, so you will have to take your salary." The next day, the dean of the school told me the same thing. So, at the end of June, I received my first payment, at the fifteenth grade of the university's wage scale.

To my surprise, Ya'nan showed up one Sunday with her bedding roll and all her suitcases. The party leadership at the feed lot had sent her back to Hangzhou to work at Zhejiang University. On Monday morning, I accompanied her to the school, and she was assigned to the bookstore. That afternoon, she went to register for a resident's card at the local police station. Much later, Ya'nan told me that she had left the feed lot because everyone knew that she was a rightist's daughter. In Hangzhou, however, her fortune did not noticeably improve; her new job included lifting heavy boxes and pushing carts. I felt extremely guilty that no one cared for her, all because she was my daughter.

On the eve of the Chinese New Year, 1961, a leader brought some fish to my house and said, "The provincial leadership wanted you to hear the good news tonight so that you will have a happy New Year, but it will be announced officially tomorrow in the newspapers. You are no longer classified as a rightist! But please excuse me now, for I have quite a few families to visit!" The next morning, I saw in the *Zhejiang Daily* the news that my rightist "cap" was removed. The *People's Daily* never did print it, though that paper had been the first to report on me as a rightist. I was so happy that I felt as if I had been reborn.

One Sunday morning, I heard someone out in the yard asking about me. When I peeked out, I saw my old friend Wang Yunru with ten eggs wrapped in a handkerchief. She came in, put the eggs in an empty pot, and said, "These are from our own chicken." We chatted a while in my little room, and she invited me to her house for dinner. Hesitating, I accepted, and she said she would have the doorman wait for me at the gate so that I would have no problem entering. When I arrived, she invited me upstairs to talk. Her daughter, whom I had not seen in many years, happened to be in Hangzhou at that time. Even though I was extremely grateful for their hospitality, I said, "Because of my background as a rightist, I do not think we should keep

in touch. I am afraid I might cause you some trouble."
There was a long silence. After the encounter, we no longer
associated with each other.

Later, the provincial Propaganda Department urged me
to take part in the Four Clean-ups campaign* in the countryside. They brought me up to date on the situation in the
outlying districts and recommended that I go there, since I
was familiar with those areas. Knowing that this work
would be beneficial for me, I opted for the village most
difficult to reach. But the department refused, since the
village was on the top of a mountain and unsuited to my
poor health. They suggested that I go to another village,
Lower Manjuelong. When I was preparing to leave, I received an offer to participate in a study group organized
by the National Cultural Association. I was one of the three
people invited from our province. Around the same time,
the leadership also called me to the library during the summer vacation to tell me that Ya'nan could attend senior
high school. She was very excited but also nervous, so she
began to review some lessons. It had been a few years since
she had attended junior high school. Entrusting her to a
female comrade, I left for the meeting.

That summer it was cool in Hangzhou, but unbelievably
hot when I arrived in Shanghai. I shared a room in a hostel
with two other people and could not sleep at all. The next
day I proceeded to Beijing, and then to the Western Hills
outside the city, where the meeting was to take place. The
All-China Federation of Literary and Art Circles put me up
in a room in the courtyard of a large house. Across from
the room of another female comrade from the Northeast,

---

*The prosecution of corruption unfolded during a political cam-
paign. Mao Zedong had called for "four clean-ups" during a central
work conference convened in May 1963 in Hangzhou itself, in the con-
text of the Socialist Education movement. The major, nationwide Four
Clean-ups campaign unfolded in 1965 in most of China; see below for
Chen's participation in it.

my tiny room had a small bed, a stool, and a desk by the window. The ladies' room was outside, but we had brought our own washbowls, which we stored under our beds. The comrade in charge of our lodging was remarkably kind and dependable. As soon as we arrived, he came to ask if we needed any help settling in. Since I had not experienced this sort of concern for years, I was deeply moved.

There were three or four other uncapped rightists in the study group.* First, Comrade Lin Mohan addressed us, and then Comrade Yang Shuo spoke about arts and literature overseas. Sometimes our meetings were plenary, and other times we broke up into small discussion groups to debate poetry, novels, or drama. The three of us from Zhejiang province were in a group that was led by a party member but also included a professional writer who received a monthly salary in addition to his profits from writing. I rarely spoke at the meetings, but I did mention the mistakes I had made during the antirightist campaign and my desire to rejoin the party and start writing again.

I decided to write Deng Yingchao that I was studying in Beijing and hoped to see her. The following day, she called to invite me to her house on Sunday morning. I informed all the leaders inside and outside the group of my intentions, and after breakfast on Sunday, a female comrade kindly escorted me to the State Council.† I entered the gate and saw Deng waiting outside the door of her house. Once we were inside her sitting room, we embraced each other tightly and I broke into tears. Also very moved, Deng said, "Please keep on good terms with the provincial Party Committee and try to patch up your status problems!" She then offered me breakfast, but only peeled an apple for me

---

*Uncapped rightists are still "ex-rightists"; cf. English usage, "ex-convict."

†The highest organ in the Chinese government apparatus, under the premier (Zhou Enlai, Deng's husband).

because I had already eaten. I was not able to see Premier Zhou, for he had not gone to bed until four o'clock in the morning, but thinking I would have other chances to talk with him, I did not worry. Little did I know that I would never see him again. Deng and I discussed everything that had happened since we had last seen each other, and she reiterated that to rejoin the party, I should stay on good terms with my provincial Party Committee.* She then accompanied me by car to Zhou Yang's house and said, "I must return home now, so please give my regards to him."

Comrade Zhou Yang and his wife Su Lingyang were at home with their daughter and grandson, and they all greeted me enthusiastically. From our conversation, I learned that during the antirightist campaign, Premier Zhou had asked Zhou Yang to entreat the Zhejiang provincial Party Committee not to classify me as a rightist. The premier had believed that a little criticism was all that was necessary, since I was merely a straightforward person who always spoke her mind and did not have ill intentions. Zhou Yang had called long distance and written to the Zhejiang Committee but had not received a response at first. Later, a leading comrade had responded: "If I do not classify Chen Xuezhao as a rightist, I will lose my position." It took me an exceptionally long time to understand and accept that the leader had been in a quandary. After we ate lunch, I returned to the Western Hills.

While in Beijing, I also visited Kong Dezhi and her husband Mao Dun, and Chen Xuanzhao and her husband Wu Juenong. Comrades Lu Dingyi and Yan Weibing were vacationing at the Beidaihe seashore, so I wrote to them later. At the end of my stay, Comrade Yang Hansheng came to

---

*It is a paradox that despite her pull at Party Central, even in post-Mao times Chen Xuezhao has had difficulties with her provincial Party Committee—partly *because* of her connections at the top, which retains her personal dossier, she indicated in 1990. The party is to this extent decentralized.

photograph the study group. I returned to Hangzhou before National Day, 1961.

My spirits were high in those days, so to expedite my reaffiliation with the party, I decided to study carefully the works of Marx, Lenin, and Chairman Mao, reform myself, and get on the good side of my provincial Party Committee. After all my travels, I finally had time to think, but there were still many things in my life that I did not understand. Since I was almost always in the countryside and therefore rarely saw any of the provincial leaders, I wondered what I possibly could have done to offend them. I never bragged or fawned upon them by secretly denouncing others. I finally thought of C. and the former party branch secretary of Zhejiang University, Xu Zicai, since they had collaborated on the letter accusing me. During the land reform, I had opposed forcing the farmers to join cooperatives and had said to C., "If they desire to form cooperative groups, let them use their own methods. A little trial and error will not harm the country." C. must have relayed her version of my statement to the leadership. But in "Tea Growers Along the Cooperative Road," published on December 21, 1955, in the *People's Daily*, I had only praise for the cooperative movement.

A few days later, I went to Lower Manjuelong. Carrying my bedding roll and a small suitcase, I took the bus to Siyanjing and walked to the village. The farmers had almost finished harvesting the sweetly scented osmanthus found on both sides of the road. The principal crops of the area, osmanthus and chestnuts, brought a fortune to the farmers. Tea was also grown, but not in great quantities. At the end of the road, there were many tourists sampling a bowl of sweet osmanthus soup at the Osmanthus Pavilion. I turned at the pavilion and continued to the village on a cobblestone road lined with houses. There were many hills and knolls in this rolling countryside.

When I found Comrade Zuo, the party branch secretary,

he took me across the street and up a hill to the house of the peasant family with whom I would be living. Only the mother and her young son remained in the house. The two daughters were married, one to a neighbor and the other to someone in the next village. The dining room, which contained a square table surrounded by stools, was located just inside the front door. Upstairs there was a comparatively formal room with crystal windows that looked out on the road to the south. Next to this room was my little room, a garret that was light because it faced north. Since there was only a frame, not a door, and there were many cracks in my wall, it was always freezing in my room, even with the windows closed tightly. I tried to cover the cracks with newspapers, but the wind still penetrated. After several nights, I contracted a terrible cough that became chronic and worsened in the winter. I had to sleep with a scarf wrapped around my neck. Finally, I went home to retrieve my mosquito net and a sheet. I hung the net over my bed and the sheet in the door frame, but I still coughed a lot when the wind blew.

Contributing my food and oil ration tickets to their household, I always ate with the peasants. The family's vegetable garden and two huge vats of manure were right outside my window. Since manure was extremely scarce, each family had to donate a certain portion of their supply to the commune. The young boy was only home for meals because he managed the storage area for the commune. When I became accustomed to their life-style, I started to help the mother make the fire, cook meals, and wash the manure buckets, since I could not handle working in the fields up on the mountain. After I had finished my chores during the day, I would call on various families to gather information. One old woman told me many tales of the past and the present. Her grandson, a Youth League member, never spoke to me but appeared extremely mature.

It was the agricultural slack season. In the daytime, the

tea growers chopped firewood on the mountain or mended the tea tents. At night, many meetings were held in classrooms at the primary school. Secretary Zuo was very kind to me and always personally invited me to them. There were party and Youth League meetings, women's meetings, and poor and lower-middle peasants' association meetings, to name a few. Often they were so crowded that latecomers had to stand out in the hall and elbow their way in when they wished to speak. I usually sat on a bench with other women and listened attentively to the discussion. After the meetings, I would return to the peasants' house with a flashlight and use my suitcase as a desk to record the most important comments in my diary.

From my conversations with the peasants, I concluded that Secretary Zuo was well-liked by the people. Although some thought he was too timid, everyone supported him. Not a native of the village, he had been a member of the organization committee before just recently becoming the branch secretary. As a boy, he was poor and often came to the village for work during the busy crop season. During the land reform, he had registered for a residence here and was given a small, two-story house with his mother. He later married and had two children.

The villages of Upper and Lower Manjuelong, and Siyanjing had come together to form a production brigade with a party branch. The branch secretary was a peasant from Upper Manjuelong. He and his older brother had worked and begged for food and slept on farmers' firewood piles until the land reform, when they both married and were given houses. When he was appointed party secretary, his brother became the head of the production brigade and also of the smaller production team in his own village. The deputy secretary of the party branch and the party propaganda director were from Siyanjing. The upper and lower villages both grew a great deal of osmanthus, but Siyanjing had very little.

In earlier days, the farmers soaked the harvested osmanthus in plum juice, which was extremely expensive but maintained the plant's original color. They then sold the high-grade product to the state at a fixed price. Private retailers always offered to buy it from the farmers at a higher price to make a profit. Now, however, people substituted a less expensive chemical for plum juice and some profiteers from Shanghai came to buy the crop. At a meeting to vote on whether to continue this practice, the branch secretary and his brother supported selling the more cheaply made product to the capitalists, because they said it improved the farmers' lives. The deputy secretary from Siyanjing was opposed, but since his village did not produce much osmanthus, the decision was not his. The proposal was approved, and the brothers made a fortune by selling a small amount to the state and the rest to private retailers. They made their trips to Shanghai by airplane and took out a room in the high-rise Park Hotel as their office, with a plaque on the door. They even had peasants move the osmanthus in their suitcases, to keep their shipments inconspicuous. But the poor peasants did not make a cent off it. The two brothers spent the money eating, drinking, and gambling at mah-jongg nightly in a temple on the mountain used for storage. They hired a landlord's widow to take care of the temple and occasionally gave her some leftover food, for which she was extremely grateful.

The people were unaware of this corruption until one day at sunset when two farmers walked down the mountain and stopped to rest in front of the temple. They were curious to hear noises from inside the building, so they entered through the unlocked door. There they were quite surprised to find the secretary, his brother, and a few others laughing heartily and sitting around a table loaded with chicken, duck, pork, fish, and liquor. When the tea growers were enlightened about this depravity, some poor peasants and hired laborers among them reported the brothers' actions to higher leadership. Little did they know that the brothers had already distributed the profits among selected members of the production brigade. Some people had been given three or four hundred yuan, others as much as a thousand. The brothers had already become agents for the landlords and criminals.

When I came to Lower Manjuelong, the party secretary had been fired from his position, expelled from the party, and sentenced to a labor farm for two years. But his brother was still the head of the production brigade and used his connections to have the ex-secretary released after only six months of his term. At that time, the main topic of local meetings was the uncovering of the osmanthus deals, because a throng of people were involved. Asserting that she needed to see a doctor, the culprits had sent the widow who took care of their hideaway to a relative's house in a nearby city. The peasants dispatched people to retrieve her, and under their guidance she divulged much essential information.

While people were preparing to celebrate the Chinese New Year, the Four Clean-ups campaign was coming to a close. I went home to Hangzhou for the holidays, but bad weather and pressing business prevented me from returning to the village until the second tea season. From then on, I only visited the village occasionally to complete my experience in the area. Eventually I wrote the sequel to *Spring Camellia* based on these trips.

By the time Ya'nan graduated from Learn from the Army High School, she had joined the Communist Youth League. One of her teachers asked me if she would like to work in the countryside, but since my health was poor and she was my only daughter, no one from her school forced me to send her away. The new director of our library, however, said, "If you send her to the country, she may return for college in two years. Your relationship with the party will be harmed if you prevent her from going!" I remained undecided, but Ya'nan, encouraged by her girlfriends, insisted upon going. So in 1964, she went away to labor in Yangzhou Commune, fifty miles upcountry in Tonglu County.

In late autumn of the same year, I read an article about someone named Xu Zicai. He had gone to Holland as part of an industrial technology delegation but he was secretly affiliated with the Soviet Union. Not knowing that people in his delegation suspected him, he planned an escape. He told his Russian friends to wait under his window, but he seriously injured himself when he jumped, and the Russians fled. Our people sent him to a hospital in Holland, but he did not survive. I was unsure whether he was the same Xu Zicai at Zhejiang University who had written the letter accusing me, until a comrade confirmed that he was. Because of his terrible misfortune, I could not be happy that he was caught, but I was convinced that his fate had been determined by his own character. Time will always tell. In return for his letter, he had obtained the leaders' trust and respect, a trip to Russia to study to be a professor, and a prestigious teaching post in Harbin. Unworthy of the party's trust, he had betrayed his fatherland. During our most trying economic times, the Soviet Union had caused major setbacks by reneging on contracts and calling back the experts they had sent to China. The case of Xu Zicai taught us that we must take care of our own experts and not rely on other countries.

At the suggestion of the provincial party Propaganda Department, I went to help with the Four Clean-ups campaign at the Hangzhou Brocade Mill. Because of my political status as an ex-rightist and my limited skill, I could only assist with simple tasks and attend some meetings. The mill was close to where I lived, but my landlord insisted that I move out because he needed the space. In January 1965, I finally found three small rooms in Hedong Dormitory. That was not as close to the mill as my previous lodging, so I only went to work in the mornings. At the beginning of the movement, every worker did a self-criticism. I later joined an advanced group that had an extremely tight schedule, and I began to write a short novel entitled *Zhang Jingxin and His Group of Exemplars*. Since I did only trivial chores and did not become familiar with the workers in the designated "advanced group" who were my subject, I was never satisfied with this book. I threw it in the trash. Later, during the Cultural Revolution, my diary was confiscated and returned with only a few pages about my contacts with the group remaining. Therefore, it was impossible for me to rewrite the book.

After I had moved, I wrote to Deng Yingchao to inform her of my new address. I received a response written by her secretary that encouraged me to rejoin the party. I had a habit of burning all letters from friends or leaders after reading them twice, but I accidentally left her letter in the book I was reading.

During the Chinese New Year holidays, Ya'nan came home. She did not mention it, but she had received a certificate of merit from her production brigade. I heard of this accomplishment from a teacher sent by the brigade to congratulate me on my daughter's achievement. The money she received at the end of the year, however, was quite insufficient to live on. I asked the director of the library if Ya'nan could return and apply to college, but a party branch secretary at the school told the director: "Now that she's already been sent down, it's out of the question for her to come back." Later, when I called on my friend the vice-president of the university, he sincerely promised that he would find a way for her to study there. But soon after that, disaster struck.

Chapter 5

I REPORTED to the mill every morning until May, when the director of the library, where I studied every day but Sunday, told me not to go any more. At the library, we read the newspapers and had small group discussions in the morning. In the afternoon, I helped sort the papers. In November 1965, Yao Wenyuan published an article about Wu Han's play *Hai Rui Dismissed from Office*.* Because of the many rumors circulating about the Cultural Revolution, New Year's Day passed in an extremely tense atmosphere.

On May 3, 1966, the *People's Daily* published an article from the magazine *Red Flag* entitled "An Epochal Event: Peasants, Workers, and Soldiers Participate in Academic Criticism." From then on, there were articles or editorials on the same subject almost every day in the newspapers. At the university, there were already some large posters denouncing a vice-president who was a friend of Deng Tuo,† a major object of the criticism. The vice-president was the one who had promised to help Ya'nan return to her studies. He was removed from office shortly after the poster was printed, so I could not rely on his assistance.

---

*The attack on the play by Yao (later labeled as a member of the Gang of Four) is considered by many as the opening salvo in the Cultural Revolution.

†Founding editor of the *People's Daily*. See *Chinese Law and Government* 16,4 (Winter 1983–84).

Next was the campaign against the "four olds."* As I was finishing lunch one day, three people, two of them from the physics department, came to search my room. After examining my bookshelves, they took away *And Quiet Flows the Don* by Sholokhov and sealed the shelves and even a locked chest of Ya'nan's. They then told me to go outside. Since I was dressed casually in short pants, I asked if I could first change into a more suitable pair of long pants. I made this request not because I was old-fashioned and thought women should hide their skin, but because I did not want to be ridiculed in my short pants. But they denied my plea and forced me downstairs, where a middle-aged man was waiting. They tied a band saying "rightist" on my left arm and paraded me first to the school, then to another dormitory, and finally back to my room. In the feudal society, when someone was about to be executed, he or she had to be publicly exposed and made notorious to teach the people a lesson. When I returned to my room, I barely had time to change my clothes before I had to be at the library. When I went out the door of our dormitory, some people holding a meeting downstairs saw me and chased me with scissors to cut my hair. As I ran away, I cried, "The party did not tell you to do this!" They followed me to the gate of the school, through which they could not pass. The leader of the so-called rebel faction† in the library was a party member from the circulation group and had previously worked for the Kuomintang. Two printing workers were also rebels. Some people sympathized with me and often related to me news that I would not have otherwise heard.

After many more articles and editorials, the Beijing Party Committee was reorganized on June 4. There were posters

---

*These four anathemas were old thoughts, culture, habits, and customs.

†The extreme radical and disruptive party in the Cultural Revolution, later said to be directed by the Gang of Four.

and slogans everywhere, saying "Smash the Liu Shaoqi–Deng Xiaoping Line!" and "Down With Peng, Luo, Lu, and Yang!"* It was one catastrophe after another. When the first propaganda team of soldiers and workers came to the school, two members came to the library and took control of everything jointly with the rebels.

We had ceased our regular work so we could devote all our time to studying the newspapers. One day at noon, a female colleague told me that I might be interested in a poster in the mathematics department. When I finally found the department, which I had never seen, I read the poster criticizing Xu Ruiyun, the department's head. It asserted that since Xu was a friend of mine, she should be classified as a rightist. For some reason, I did not mind the poster, probably because it was signed and seemed relatively honest. I first met Xu at a meeting. Childless, she became attached to Ya'nan and often helped her with her math. When she was younger, Xu had suffered from tuberculosis, and her parents had not paid much attention to her since she was a feeble girl. She spent most of her time studying at her house and eventually became an accomplished mathematician, receiving a doctorate in Germany and later publishing books. She was an extremely talented professor and loved to instill her methods in the new, young teachers. Xu had joined the party in 1956 and, like me, was a simple and straightforward intellectual without practical experience in the community. When I was classified as a rightist, she came to me to discuss her lack of understanding of many things in our society. I told her to forget about me and never to contact me again. We had not seen each other since then. I once saw her walking toward town in a cheongsam, a traditional Chinese dress, but went another way to avoid encountering her. I guessed that she

---

*Peng Zhen, Luo Ruiqing, Lu Dingyi, and Yang Shangkun ("The Four-Family Tavern").

was going to receive some foreign visitors but wondered why her colleagues had not driven her into town in an automobile.

Ever since the time in 1945 when I was given permission to visit France and then suddenly I saw it rescinded, an "unhealthy" thought kept occurring to me, that the feudal traditions of patronage and of ousting people from their positions because of jealousy were still in practice. Everyone fabricated tales so they could get ahead by denouncing others. Wasting its energy on such deceptions, the country had not developed but instead had fallen behind, and no one cared enough to ameliorate the situation. At this point I expressed these incorrect views of mine, begging the party to set me straight.

That day when I arrived at my room, a teaching assistant who lived on the third floor of the dormitory and was a leader of the rebels came to see me. He ordered me to move everything from my kitchen and dining room into the hallway and then broke my stove by smashing it down on the cement floor. Since the factories had already shut down, "refusing to do productive labor so long as the general line was incorrect," I knew it would be extremely difficult to buy another stove. He then told me that unless I moved out, he would throw everything into the street. Even though I was sixty years old, ill, and classified as a rightist, I was prepared to sleep on the street. The man insulted me at the top of his lungs: "Kuomintang! Traitor! Collaborator! Spy!" I thought, pity that I've never done any of that!

My maid at that time was from a poor but honest peasant family. Her husband was a rickshaw driver, and since she lived far from my dormitory, I gave her additional money to ride the bus. She brought a bag lunch each day, and in the afternoon she either mended my old clothes or rested in Ya'nan's room. The teaching assistant ordered her to go home: "You are not allowed to work for a rightist!"

My maid replied, "I do not care that she is a rightist, and you cannot tell me to leave. I live on my own work; I am a laborer!" The rebel could not coerce her, so she helped me move more things into the hallway. We had nowhere to cook. The rebel then tacked a poster outside my door requiring that I clean the public toilets every night after ten o'clock.

A female colleague from the library warned me that the rebels had commenced raiding people's houses and confiscating their possessions. I did not know how to prepare for their arrival, so when I saw them coming, I put my bank statements in my maid's pocket. The rebels looked around and seized all my books, including my *Sibu Collection* and *Kangxi Dictionary*. I only had enough time to record how many books were taken, not the individual titles, so I tied them into separate parcels and gave the rebels a sheet listing the number of books in each parcel. When I asked them for a receipt, they promised to give me one later. I was too naive and trusting to insist upon getting one at that moment. A few days later, I inquired about my receipt. I was told that my inventory sheet had been lost, and the rebels had forgotten how many books they had taken.

After taking all my cash from a drawer, the rebels said, "You must have some money in the bank. Where are the account statements? If you do not give them to us, we will search until we find them!" I was afraid of involving my maid, so I took the papers from her pocket and handed them to the rebels. One of the men commanded me to send her home: "You are not allowed to have a maid." But then someone from the local police station instructed her to keep me under close surveillance, so she was not allowed to leave.

Later, my maid told me that a rebel had pointed at two jars of sugar and said, "Look how much sugar she has. Do we workers have this?" When my maid declared that everyone had the same amount of sugar, he asked, "What sort of medicines and tonics does she use?" My maid told him that I never used them, and he replied, "You should denounce her!" She retorted, "I have no reason to accuse or to expose her. She is very humble and treats me quite well." Knowing that he had not succeeded in turning my maid against me, the rebel angrily told her to go away.

I did not have a penny to my name and did not know what to do about the maid, because the local police had ordered me to keep her. She offered to work all day for only half a day's wages and said that if I could not afford that, she would lend me some money. I knew I would never forget this kindness and sincerity. When no one was watching, she cleaned the public toilets for me, which were extremely dirty, especially the men's. The school knew how to collect rent, but it never used the income to repair the toilets or dormitories. It was much easier for me to clean the toilets after my maid had disinfected them at least once. Because of my heart trouble, I became quite weak and my heart beat rapidly when I exerted myself scrubbing.

When I returned from the library one day, my maid removed from the trash can a piece of folded paper she had found in my room when she arrived. She handed it to me. It was an unsigned poster. The last two lines read: "You are a little running dog of Liu Shaoqi and Zhou Yang. We will never let up until your body is crushed and in pieces!" I was shocked at these words and agreed with the maid that the neighborhood children whose parents were rebels had probably printed the poster.

Since my residential area belonged to the school, the rebels in the school governed everything there. They ordered a group of seven former rightists and capitalists to pull weeds in a large yard at four o'clock each morning. Even though I was the only woman, the younger, stronger men were still harsh with me and left me the most difficult work. I simply ignored them and worked by myself in a corner. A few days later, everyone in the group except for a rightist from the physics department and me was taken away for investigation. My new job was to clean the streets, at the same hour in the morning. It was winter, but after cleaning awhile each morning, I warmed up and removed my cotton jacket. I started suffering spasms of pain in the nerves of my hips and thighs. This pain became more and more acute each day and, at best, I could sleep only four hours a day. I realized that they seriously planned to destroy my health!

I do not remember how many times the rebels raided my house and seized my possessions. They had long since taken my diary, so I could not record any events. One of the rebel leaders always carried a bamboo whip and waved it in my face to threaten me. Once he said, "You may no longer read French books!" He then upset all my boxes and chests, checked all my pockets, and confiscated all my French books. As he left, he motioned at me with his whip, ready to use it at any moment, and said, "If you still have any French books, beware!"

When the rebels unexpectedly arrived to examine my desk drawers, I found I had been too careless. They found a photo album and a draft of my career resumé in which I had recorded important dates and other information. When they tried to seize a photograph taken in 1955 of Deng Yingchao and me, I could no longer tolerate them and asked, "What is wrong with her?" I refused to part with the picture, so they continued to search and found the letter from Deng that I had left in a book. They took the letter and asked me many questions about Deng, to all of which I replied, "I don't know." When a female rebel said she had heard that Mme. Deng was suffering from tuberculosis and asked me about it, I said, "I don't know." I never liked to discuss personal problems, even with old friends.

I had already begun study classes with my fellow "cow

demons and snake spirits," the nickname given to the rebels' class enemies. I was lumped together with enemies from the library, the physical education department, and the maintenance staff. The rebels in those three areas directed us. The customary Lin Biao–style morning service to ask for instructions was altered to suit us. Since we had already erred, we confessed our faults and asked for punishment. We knelt in front of a portrait of Chairman Mao in a hallway of the library, held up the Little Red Book of Mao's quotations in our right hands, and chanted loudly, "With full respect, we wish Chairman Mao and Vice-Chairman Lin Biao eternal life and good health." When we adjourned to a larger room, I was the only one who was interrogated. The rebels kept asking me: "What did you do before you came to Zhejiang Province? What did you do before 1957?" I usually remained silent but sometimes replied, "What did *you* do previously?" I always felt depressed after these inquiries.

During study time one day, I went downstairs to the toilet and saw Jiang, the rebel in charge of us for the day. In a low voice, I asked him what I should do when they questioned me about my life before I came to Zhejiang. He replied, "Say absolutely nothing! Talk only about the anti-rightist campaign." His simple words inspired me, so when I was interrogated later, I followed his advice. On one of the extremely rare evenings when there was not a meeting to criticize me, Jiang came to see me after I had returned from my study. He reassured me: "Relax and sleep well. Do not worry about anything." He then checked my door and said, "It is not secure. People could easily break in. You must put on some bolts, not one but several. If anyone knocks at night, don't answer until other people wake up and come out." Even though he said the words in a harsh tone, I realized that he was genuinely concerned. I told Xiquan to buy some bolts for my door, and her oldest son came to install them for me.

The rebels' next assault was on a mathematics professor whose brother was considered a revolutionary martyr since he had been killed in the War to Resist U.S. Aggression and Aid Korea. The rebels hung a huge poster board around his neck asserting that his brother had not died honorably, and that his brother-in-law was a Kuomintang member who had gone to Taiwan. The poster was so large that he could not fit through the door. The rebels paraded him around to expose him to the public. When he returned home, a rebel from the foreign languages department came to tell him to move out. The professor was forced to obey and moved his family to a room that had been used as a kitchen. They moved their stove to another kitchen that they shared with another family. Neither he, his wife, nor his two daughters went out in public for three days. It was said that the rebels searched his rooms three times to find some letters from the sister who was married to the man they denounced as a Kuomintang member. I had a hard time understanding the rebels' behavior, because parading people around, cutting their hair, and forcing them to kneel were feudal tortures used two thousand years ago. I did not see how a socialist country could restore feudalism to such a degree!

The rebels often seized me to exhibit me to the people. Once I was struggled against together with leading comrades in the provincial party propaganda apparatus. The rebels took me to a meeting hall and ordered me to stand on a bench that was on a narrow table on the stage. The audience jeered while the rebels on stage criticized me, and some young Red Guards behind me threw stones at my head. I remained composed as I made the difficult climb up to the bench. I came out of the meeting with a few lumps on my head from the stones, but afterward I realized that if I had fallen, I could have been killed. Neighborhood children knew me now, so every day on my way to school, a group of them threw stones at me. One day as I was walking home, I saw Comrade Fang Lingru, another object of criticism, getting off a bus. When she saw me, her eyes filled with tears. Since I feared the others were listening to us, I simply said, "We have to keep on living!"

Classes were suspended since all the elementary, secondary, and college students were rebelling against their teachers, whom they labeled as capitalists, reactionaries, or "stinking ninth elements,"* an insulting name for the intellectuals. The more knowledge one had, the more reactionary one was. Since they did not want to produce anything for the wrong people, the factories were closed altogether. The rebels had attacked both military and domestic areas, seizing and destroying things everywhere. Twice as I walked to school, some young men tried to steal my watch, so for several years I did not wear one, especially in the summer.

We "cow demons" in the library were now separated from those in the other departments. The head of the library and I were the principal objects of denunciation and were therefore treated very poorly. The rebels continued to ask about my activities before I came to Zhejiang, and one day they beat the library head and wounded his shoulder so badly that he could not lift his arm.

Two years in a row, Ya'nan did not come home for Chinese New Year. She never told me about the hardships in her life. I later found out that her room in the countryside had been raided many times. Among the young people from the city were some rebels who called Ya'nan "little rightist" and took away her bed. As a result, she had to sleep on the wet, muddy floor, which made her back extremely sore. She did, however, meet some kind people. A poor peasant's daughter named Lianxiu invited her to

---

*The Four Black (bad) categories were expanded to five, to seven, and finally, in the depths of the Cultural Revolution, to nine, to include all intellectuals as a group in the last category.

sleep at her house. The situation in the countryside was dreadful. No one worked in the fields, so all winter long the only available food was sweet potatoes. As the rebels had destroyed all the vegetable gardens, people had to dig for wild herbs to eat. I had previously sent Ya'nan twelve yuan a month for expenses but now was no longer able to, since I was in debt to Mo Xiquan and my maid. The school paid me forty yuan a month at first and then sixty yuan—much better than the salaries of some other "cow demons," which were only twenty yuan a month. But with my medical bills, this was never enough. I later began to send Ya'nan money again, usually about ten yuan a month. Xiquan took some of my old clothes and my empty bookshelves to a second-hand store, but since I was not the only intellectual to do such a thing, she did not receive much money for them. I had to keep my piano in the hallway because it would not fit in my small room, and my rebel neighbors used it as a table. Often they would put hot dishes on top of it. Finally the Zhejiang Song and Dance Company convinced me to sell it to them at a low price.

Since my right elbow had been dislocated several times, the young men on the street always yanked on my right arm in addition to throwing stones at me. One day they succeeded in dislocating my elbow again, and I went to the hospital; I could not even lift my arm. The doctors often refused to treat rightists, but luckily they did not discriminate against me. The acupuncturist who treated the pains in my hip was an amateur. She gave her best effort to finding the correct points but never hit the exact spots, so I transferred to an experienced doctor in another hospital. He was extremely helpful and compassionate, and his fees were always reasonable. In those days, the pain was so intense that I could not lie comfortably on my bed. I asked for sick leave from work, but the rebels still dragged me to every accusatory meeting. I walked slowly with the help of a stick to those meetings. One time, the rebels took away my stick, and I was forced to kneel down since I was unable to stand. When they adjourned for supper, no one assisted me in standing, and one rebel shouted, "Crawl home!"

After a year and a half of acupuncture, my right leg was no longer in pain, but I found out that I had muscular atrophy. My doctor promised, "Your treatment is finished, your leg will not bother you any more." Since no one believed that I was afflicted with a disease, I still studied every day and cleaned the streets each morning and the bathrooms each night. The rebels thought that I was pretending, but all my doctors genuinely sympathized with me. They couldn't understand what crimes could warrant such cruelty by the rebels. Not trusting anyone, I told them not to think me so innocent, because I had indeed made some mistakes.

One morning, two rebels and their leader with his bamboo whip called me out of the study room to be interrogated in a large room. They ordered me to sit down and said, "Tell us everything you know about Premier Zhou. You will be held here and not allowed to go home if you do not speak!"

I gave my stock answer: "I don't know."

"You don't know!" the leader shouted. Punching my head, he asked, "Do you have a brain?" When I repeated that I knew nothing, he hit me again. I was silent until the rebels left for lunch, when the leader said, "Think it over!" After they left, I sat up straight in my chair and wondered if they were going to starve me to death. Since I had been on the verge of death two times before, I was not disturbed, but I did resolve not to cause any trouble. They returned at two o'clock and asked, "Have you decided to obey?"

I replied, "I know nothing and that is the truth. I cannot lie!"

Enraged, the leader slammed his hand down on my

table. Then the rebels left for another two or three hours. When they came back, they told me to go away. I stood up, gathered my belongings, and left. I did not arrive home until at least half past five.

The rebels' new rule was that the "cow demons" could only enter the library through a small side door, not through the front door. One morning as I entered by the side door, I saw a slogan that read: "Down with the incorrigible capitalist-roader Xu Ruiyun!" At the time I did not realize that it had been posted especially for me.

One day in January, a colleague informed me during a study break that Xu Ruiyun had died. As time went by, I pieced together the story with the fragments that I heard. The rebels had separated Xu Ruiyun and her husband to interrogate them individually. They forced Xu's maid, whom she had always treated as a sister, to write a poster condemning her, but the poster did not say anything of substance. The rebels also forced her niece, whom she had raised as her own child, to denounce her with a poster, and they moved her from Xu's house to a dormitory. One day, the rebels made Xu Ruiyun sit through a meeting to accuse Professor Chen Jiangong, but she did not know that it was the last of the meetings against him and that he would be going home soon. After she returned from the meeting around noon, a kind female neighbor came to offer her some lunch. But when the neighbor opened the door, she saw Xu Ruiyun hanging from the door frame and fainted. Because of her high blood pressure, the neighbor also died that very night.

During the long months when I lay on my bed in pain, every time I looked at the door frame I thought of Comrade Xu Ruiyun. And I also contemplated suicide. Dying was not easy, but living was even harder. Each day was a struggle against the natural environment, the class system, and oneself. Those in the party who wore hats to designate that they were taking the capitalist road were to be eliminated. Although Chen Jiangong was freed, he was still considered a "stinking ninth element." When he was ill, the doctors refused to administer medicine to him, so his wife had to give him injections herself. But she had no experience, and all the medicine dripped out of the syringe. When he was in critical condition, he was left out in the open, and no one took care of him. Not daring to show compassion for a capitalist-roader, everyone who saw him quickly walked away, and his death went unnoticed by the people.

One morning late in March 1969, a colleague came to the room where I was isolated for study and asked me to accompany him to the meeting hall for an announcement. He told me that after the pronouncement was read out, I should shout something like: "Long live the Communist Party! Long live Chairman Mao!" When everyone at the meeting except me was seated, the leader disclosed the party's decision to liberate me and said that he hoped I would be a dedicated worker. After these two simple sentences, I cried, "Thanks to the party and to Chairman Mao! Long live the party and Chairman Mao!" I was then dismissed so that they could continue the meeting.

After this meeting, a member of the Workers' Propaganda Team revealed to me that they had gone to the Central Party Organization Department in Beijing to consult my personal dossier, because they could not find the copy in the provincial archives, only my confession during the antirightist struggle. Since they were criticized by the Beijing comrades and denied access to my file, they appealed to Premier Zhou. He was too busy to see them. But Zhou's wife, Deng Yingchao, spoke for me instead. She told them that I had originally had no problems, but that I had been unable to maintain a friendly relationship with the provincial Party Committee. Evidently they had kept my file out of the hands of the Chinese Writers' Association, too.

Later, the head of my residence committee, the young

woman who had incited the neighbors to cut my hair off, but since been paraded around by the rebels herself, whispered to me that there had also been a meeting in our neighborhood to announce my freedom, and that the young people were instructed to stop harassing me. But I continued cleaning the dormitory toilets every night until a leader called on me in June and was shocked to hear me tell him of my taxing job. He did not understand that as a notorious rightist I had already been degraded and deprived of my freedom of speech. And everyone loved the clean toilets. I also still cleaned the floors at the library, but I was not allowed to touch the rare book section, probably because the rebels did not want me to see that it was filled with my own books.

The big-character wall posters that often degraded the names of innocent people had a vast influence upon the public. As Lin Biao said, "A lie repeated a thousand times becomes the truth." At first I was curious as to what the posters said, but after reading a few, I found no concrete or reliable information. There was even one poster condemning a doctor for hiding underground Communist Party members before the takeover in 1949. The enemy within the party, the "Gang of Four," were now able to go through with what the Kuomintang never was. So maybe it made sense after all.

In November 1969, Lin Biao ordered the entire school to move to Mogan Mountain, because a Soviet invasion seemed imminent.* Ya'nan helped me transport my belongings to the top of the mountain, where most library personnel were relocated. All the women in the library and the biology department lived in a single room with only one bed. Another older woman and I shared the only bed, and everyone else slept in ranks like sardines on the floor.

---

*Armed Sino-Soviet border clashes began in March 1969, at Zhenbao (Damansky) Island and elsewhere.

The room was in a confiscated capitalist's summer house that the leadership had borrowed from the local township. Our main work was to take turns preparing meals. Firewood was a problem, because the cart we used to haul it would only go halfway up the mountain. From there, we transferred the wood to our backs and walked up to our lodge. We envied the people living at the bottom of the mountain, because they could simply walk down the street for a snack in a restaurant or pastry shop.

I was not delegated to carry wood but felt uneasy watching the others straining themselves. The one time I volunteered to help, I perspired a great deal and had to remove my coat. That evening, I came down with a high fever and was too ill to get out of bed even when the doctor arrived the next morning. The medicine he gave me did not seem to help, so he sent me back to Hangzhou on the third day of my fever and wired Ya'nan that I was coming. Since there was nothing to eat at home, we had to borrow some vegetables from the neighbors. While I was at the mountain, my maid had returned home to help her son with the chores. She could no longer work for me, so the wife of one of the men working in the library volunteered to assist me, for she and her husband were having trouble making ends meet. Helping with the shopping and cooking, she worked four to five hours a day and went home for lunch. After a month, I felt much better and could walk around. Ya'nan was at home with me, but she was always having problems with her stomach.

One afternoon, the party member in charge of the caretaker contingent of workers in the library visited me. He sat by my bed and said, "No man is going to marry Ya'nan since she is working in the countryside and is the daughter of a rightist. Young X. does not have a girlfriend, and they would make a good match. But since his salary is low, you must give him thirty yuan a month if they are to get by. You must also know that X. will never be able to join the party. His father used to be the boss of a rickshaw company."

"My daughter was brought up by the party," I replied, "and the decision is hers. I cannot move her around as if she were a piece of furniture. And I've always treated people democratically anyway."

"All right," he said. "Just propose this plan to her and if she approves, I will bring X. here tomorrow evening to meet her. If she is opposed, ask her to give me a sealed note expressing her disapproval tomorrow morning, but not to show it to anyone." He then left.

After I related the whole story to Ya'nan, we both decided that we opposed the match. X. was a rebel leader at the library, and everyone who had the least amount of sense called him a "little rogue" (*liumang*). I wrote a note saying that Ya'nan already had a boyfriend and was not interested. Ya'nan delivered the note the next morning. That evening, however, the man brought X. to my house. They sat down and asked me to call Ya'nan out of her room. Since I could not get up, the library head went to her room and said, "Your mother would like to speak to you." When she came out, X. pretended to be very shy, keeping half of his face hidden with his coat collar. Ya'nan sat on a stool by my bed and did not say a word. After three minutes of silence, she stood up and returned to her room. When they left, they seemed offended. Ya'nan and I never discussed this episode. I did not know whether X. had asked to be introduced to Ya'nan or the library head had offered him the encounter to please the little rebel leader. In any event, they hated us from then on.

Since I desperately needed Ya'nan at home to look after me, I asked my friends Wang Yunru and Zhou Qiaofeng to deliver to Premier Zhou and his wife a letter explaining my situation. I longed for my daughter to be sent back to the city. Premier Zhou finally approved my petition and said that Ya'nan could work in the university, as she had before

she went to the countryside. The leaders of the school refused, however, and replied, "Her mother has a history of
political problems, so it is unacceptable for the girl to work
here." Ya'nan was sent to another school to work in a ther-
mal treatment factory, a heavy job unsuitable for a girl. She
reported for duty in May 1971 and was a trainee for two
years.

I felt honored when one day the Shanghai People's Publishing House asked me to translate the *Memoirs of Hope* by
Charles de Gaulle. That morning I met the new leader and
decision maker of the school in a small meeting. He had
been promoted because of his outstanding performance as
a rebel. Two men from the publishing house were sitting
with him, and I was seated with a professor from the history department who had been educated at a university in
Lyons, France. The leader expressed his hope that the professor and I would translate de Gaulle's book, because it
had to be finished before President Pompidou's visit to
China. I told him that as a rightist I was not qualified for
the job. At these words, the leader said nothing, but he
reddened slightly because he was ashamed that he had
helped to classify me as a rightist. A man from the publisher said, "This is a glorious assignment! Surely you
wouldn't refuse to translate it for Chairman Mao?"

I accepted the task. I had hesitated not because I wanted
to put on airs, but because I wanted the most qualified
person to do the job. The history professor skimmed
through the book to decide which chapters he wanted to
translate and left the rest to me. I then took leave from the
library so that I could spend all day working at home. Each
evening, Ya'nan typed the draft I had finished that day.
Because French had acquired much new vocabulary since I
had learned the language, I could not find many of de
Gaulle's words in my old dictionaries. I was forced to ask
Li Xin, an old war friend living in Guangdong, to buy some
new dictionaries for me. After much difficulty, he found

two, both expensive. But, lacking many of the technical and proper nouns I needed, they were not much better than the old ones. Fortunately, the library had just bought a new edition of the *Larousse,* and I was allowed to borrow it once the leaders had certified it as acceptable. Working constantly, I translated about two thousand words each day and finished the book in a month.

The rest of the school started filtering back from Mogan Mountain. But all classes and work were still suspended. The period of rebellion was still with us. As usual, I cleaned the floors and dusted the tables in the library. I occasionally went upstairs to organize the newspapers or to help Comrade Zou cut out important articles and paste them on poster boards. Originally from Jiangxi Province, Zou had come to the school with her husband, a biology professor. During the oppressive times, she was one of the very few just people. When she had to watch us "cow demons," she followed the true party line and never forced us to do or say anything. She was extremely kind and sincere, but the party leadership at school did not pay much attention to her. Since she had been educated at a normal school, the leaders doubted her intelligence. But the leaders were not erudite themselves. For example, the propaganda officer who had attempted to introduce X. to Ya'nan had taught in the Chinese department after he graduated. But he had become notorious as a poor instructor before coming to work in the library.

Since I had started clipping articles with Zou, I had become somewhat relaxed, so the summer passed quickly. But on September 15, 1972, there was something peculiar in the air. Later the newspapers reported the story. On September 13, Lin Biao had attempted to escape from the country with his wife Ye Qun and his son Lin Liguo, but their plane had crashed in Mongolia. At first I was surprised, but when I remembered that he had been plotting to overthrow Mao, I knew it was not an isolated incident.

Following his death was the Criticize Lin Biao and Confucius movement, which was actually aimed at our beloved Premier Zhou. The rebels made many posters and pamphlets and required everyone to write a poster or say something at a meeting that concurred with the rebels' opinions. I did not want to speak, so I simply copied some slogans from the newspapers. There were often posters against me that said, "Watch out, rightist Chen Xuezhao! Do not think about returning to power!" I knew that X. had written them.

Every day when the study group was reading the newspapers or in discussion, I tried to sneak a glance at the *Reference News*, the only newspaper I could rely on.* I had decided not to read the lies, exaggerations, and empty talk in the other papers or listen to the radio, except for the weather forecast, since the only programs they ever played were Jiang Qing's model Beijing operas. By that time, I had been expelled from all artistic circles for over fifteen years and had come to realize that the Gang of Four were more of a detriment to our culture than the Antirightist campaign. The Gang, led by Mao's wife, was a dictatorship pure and simple.

In 1973, Vice-Premier Deng Xiaoping came out of retirement for a brief period, and China's situation drastically improved. The people felt much relieved. Some of my books were returned to me, including the *Sibu Collection*. But I still only had a fraction of the books that were confiscated, because many of the history books and *Kangxi Dictionary* had been stolen from the library and sold. Immediately after Deng Xiaoping reassumed control, however, the rebels launched a movement against old leaders who attempted to regain power.

At that time, a new leader for whom people had great

---

*A classified digest of translated clippings from foreign newspapers, intended for cadres' eyes only.

expectations was sent to the school. When I explained my situation to him, he caused a stir by ordering the library to search for my books. Many people in the library accused me of lying and made posters to criticize the new leader for assisting a rightist. So ended the matter. Later, my diary and some other books were returned, but as I was reading through them, a party member seized my French translation of *War and Peace* and said, "You cannot read French books!" He then ordered me to follow him downstairs, where the rebels were burning our classic books, paintings, and rare works such as Buddhist scriptures. He threw my books into the fire and said, "No more French books. It is better to burn them anyway, because they are old." I tried to grab them out of the fire, but he was stronger than I. When the school leaders later discovered what was going on, they prohibited the rebels from burning any more books. I recounted the whole story to the new leader and told him that I was still missing the letter from Deng Yingchao, the draft of my career resumé, and my photo album. The rebels denied taking these things and again accused me of lying. The man who had thrown my books into the fire put up a poster saying that I was a liar and asserted that he had not burned my books. I couldn't help laughing when I saw it.

A second group from the Workers' Propaganda Team came to the library, because the school leadership had not been fond of the first group. By that time, almost everyone in the library had gone to work at a May Seventh cadre school in the countryside. These schools were named for the day in 1966 when Chairman Mao proclaimed that all intellectuals needed practical experience in the countryside. The team was instructed to send at least one member to the countryside, but all three members excused themselves because of poor health. They remained in their little room downstairs all day and were mainly concerned with the procurement of fresh food from the country, since there was a lack of vegetables in the city.

My new task was to make catalogue cards for all the foreign magazines and newspapers in the library. I was all alone on the third floor, which was stifling in the summer and freezing in the winter. It was so hot in my tiny room that I had to take a desk out by the stairs to work. Of course, X. did not go to the countryside, so he frequently came up to the third floor to peek at the reading material. He often talked to me, but I only smiled and ignored him. One day, he said, "There is no ventilation in this room! The window above the door should be opened!" I murmured in reply but knew that I could not climb up to open it. When he came up the following day, he moved a table beneath the door and climbed up to open the window. But this act did not alter my bad feelings toward him. By that time, the Workers' Propaganda Team had left the school.

Kong Dezhi had passed away on January 27, 1971. Mao Dun, her husband, wrote to inform me a few days afterward. I sent her family ten yuan for a burial wreath, but they immediately returned my money. My sadness was unbearable, because it had been a long time since I had been in close contact with her. Only when I went to Beijing six years later did I discover that her son Sang and his wife Man had been in the countryside at the May Seventh cadre school when she died. Her grandson had taken care of her in the hospital while her little granddaughter stayed at home with Mao Dun. The urn for Dezhi's ashes was kept at their house. I could not control my tears; she should not have left this world at her age. The doctor had incorrectly diagnosed her illness, which happened quite frequently in those days. As I recalled all the times she sincerely offered to help me, the half-century of our friendship flashed through my mind as if it were yesterday. Each time Ya'nan or I mentioned her name, we both broke into tears, and I did not sleep well for at least two weeks.

Comrade Chen Yi, the former foreign minister, died of cancer on January 6, 1972. There was a photograph in the newspaper of Chairman Mao holding the hand of Zhang Qian, Chen Yi's wife. The rebels profaned his name at his memorial service and said he had carried on illegal relations with other countries. I knew that Chen would have lived longer if the Gang of Four did not exist. Many people privately felt that Chen had been wronged but did not dare express their opinions in public. A colleague told me that when in agony, Chen had said, "In the end, goodness will be rewarded, and evil will be punished. It is only a matter of time." We both agreed with his statement.

My meetings with Chen Yi had been infrequent, but each time he had made a deep impression on me. One of the few leaders who did not inspire great fear, he was extremely humble, straightforward, and easy to get along with. Although I did not know him well, he was quite familiar with my situation. He was a poet; he deeply loved and understood our culture. When he accompanied me to visit Huang Yanpei, Fu Sinian, and Zhao Chaogou, he had suggested that I write Cai Boling to persuade him to return to China. I did so, and Chen forwarded my letter to Cai. After Liberation, in the spring of 1953, Chen, Nie Rongzhen, and Li Bozhao had come to Longjing, the tea village where I was working, and we had talked and shared meals together. Chen had taken a photograph of Bozhao and me, but it was destroyed during one of the many rebel raids on my house. The last time I saw Chen was in 1955, when together we received Jean-Paul Sartre and Simone de Beauvoir. When I reflected on his poem about a pine tree remaining upright in a heavy snowfall, I understood his view of reality.

Those who had gone to the university May Seventh cadre school returned from the countryside. For the first time, peasants, workers, and soldiers were graduated from the school. They had been selected from the army and various factories and farms because of their outstanding performance as rebels. Instead of studying at school, they continued to revolt. The students had one thing in common: they were all extremely vociferous at meetings and always fabricated wild tales to denounce others. After they graduated, some were hired to work at the school, and some were assigned to the library. One rebel who had studied in the foreign languages department and recently joined the party became my immediate superior. He supervised my cataloguing of the foreign journals and sometimes attempted to help me. I had forgotten almost all the English I had learned as a child in school, but I still found hundreds of mistakes in the cards he made. I did not understand how he had passed English in the foreign languages department! He translated *Time* magazine as *The New York Times* and *Woman's Day* as *Newsweek*, so I had to redo all the cards he made, which was an enormous waste of the library's cards and my time. Later, probably from embarrassment, he left to help the circulation group.

Sad and unexpected events occurred one after another. On January 9, 1975, after devoting his whole life to the revolution, Comrade Li Fuchun passed away, resentful and downhearted.* I could not believe my ears when I heard the news. Since I had come to Zhejiang Province, I had rarely had the chance to see him. Our last meeting was in the fall of 1954. He approached me and made his customary inquiry about my well-being. As usual, I replied, "Mmm . . . " He knew me so well that there was no need to explain. I could not recall the innumerable times Li had helped me. In my worst days, he had consoled and instructed me. I tried to write some articles in his memory but failed since it was unusually difficult to portray his unique personality as an exceptional revolutionary. Every

---

*An old Hunanese friend of Mao Zedong's, Li Fuchun was one of China's earliest Communist Party members. Educated in France, with his classmate Nie Rongzhen, Li became chairman of the State Planning Commission in 1954 and, in the 1960s, China's top economic administrator.

time I thought of him, I remembered our first encounter in Yan'an in August 1938. After the fall of the Gang of Four, one of my relatives located a copy of my 1939 report, *Interviews in Yan'an*, and copied the entire book by hand for me. After rereading it, I was reminded of my conversations with Li in Yan'an and amazed at how much I had forgotten. He was always open-minded, honest, and sincere. I made it a point to reread my old reportage to cherish my memory of him and learn from his wise words. In those days, I was not allowed to express my sad feelings, but I was so upset that I had to write a letter to his wife, my Big Sister Cai. I doubted, though, that it would ever reach her.

When the pain in my hip first started, I had taken a sick leave for six months, and my salary had been reduced by 30 percent. Later, even though I had returned to work full time, the library continued to give me a reduced salary for almost two years. Since I had endured many more taxing hardships, I never worried about getting my money back. But one day, the leaders announced at a meeting that they were going to reimburse me. A long time went by, however, and nothing happened. When I finally ceased working at the school, I told them that the matter could easily be cleared up if they examined their ledgers. Asserting that the records did not go back far enough, the rebels refused to investigate. But when asked, the bookkeepers guaranteed that they still possessed all the accounts from many previous years. I was sure that someone had used the money illegally.

I could excuse the damage done to me personally, but not that done to my country. The younger generation did not care to work or study, but indulged in lying, cheating, and injuring others. At the meetings in the library, the topic was always the tremendous number of traitors and spies in the party. The more enemies found inside the party, the happier the young people were! Even though I tried to ignore them, I was never content, because my heart disease had worsened. I could only work half-days, because the doctor had found that I had angina pectoris.

Early in the morning of January 9, 1976, I heard a funeral hymn over the loudspeaker outside. Ya'nan and I burst into tears when we turned on the radio and heard that Premier Zhou had passed away the day before. My eyes were red when I arrived at work, and some party members jeered, "That old woman washes her face with tears these days!" After Zhou's death, no one was allowed to wear mourning bands or to hold services in his memory. But the people persisted, and finally a memorial service was held in the meeting hall of the school. None of the main leaders showed up, and, to show the unimportance of the service, the lesser leader who did come did not give the customary speech. But it was standing room only in the meeting hall. One volunteer after another got up on stage to speak, and after a while the speeches were drowned out by the sobs. Unable to control myself, I broke into tears. Ya'nan and I had not eaten all day, but we were not hungry. As the party members had said, I was washing my face with tears while Premier Zhou's ashes were scattered over the mountains and rivers of his fatherland. Our country had lost such a valuable proletarian revolutionary! People felt as if they had lost a father. They chanted, "We will miss you forever, dear Premier Zhou!" I dared not think about his widow Deng Yingchao and how she would survive.* I did not write to her, because I had known for a long time that all my letters were opened and read.

After the premier's death, I wrote a poem in his memory and hid it in a small notebook. One stanza went:

> I long to return to Zengjia Cliff.
> I long to return to Yangjia Hill.

---

*In fact, Deng Yingchao survived long enough to provide window dressing for the post–Beijing massacre leadership of 1990.

I long to return to Zhongnanhai,
But . . .
With countless memories and tears
I will engrave your words on my mind forever.

The year 1976 was a calamitous year, including an earthquake in Tangshan. On July 6, Marshal Zhu De passed away. I went to the library with a mourning band on my left arm. When the party branch secretary saw me, he pointed to the band and exclaimed, "You cannot wear this!"

I said simply, "I want to." Then I asked him what he had done before coming to the library, because everyone knew that he was extremely proud of his past service as a guerrilla in the Siming Mountains. As I expected, he began to recount the good old days, but I interrupted him and said, "Oh! That is wonderful! You do know Commander Zhu De, don't you?" He caught my sarcastic tone and did not bother me any more.

At the beginning of the summer, I had a problem with my skin and thought at first that it was prickly heat. I heeded Ya'nan's advice to wash my skin with hot water, but my condition only worsened. There was a red circle in the middle of each bump that prickled the most. Since I could not climb onto a bus because of the pain in my hip, I wrote to describe my symptoms to Dr. Lin Nengwu, who had cured my scalp disease. My elderly maid and my adopted daughter Mo Xiquan delivered the letter. Dr. Lin said that my skin problem was caused by contact with artificial silk and gave me some lotion to apply three times a day. I was amazed that he knew the cause, because I had not told him that I slept in such a silk shirt. I used the lotion all summer, and the itching gradually disappeared. I later found that I had actually had ringworm.

On September 9, 1976, I learned of Chairman Mao's death. The school held a formal memorial service, and everyone was extremely worried about our country's future. I often reminisced on the past and thought of the countless souls who had given their lives in furtherance of the revolution. To memorialize my friends Mai Xin and Hu Nan, I leaned up in my bed and wrote about them in "Memorable Years and Unforgettable Comrades."

Two days before National Day, I considered going to thank Dr. Lin and his wife Dr. Yu, because I felt uncomfortable simply writing letters to them to obtain my prescriptions. Not wanting to make the exhausting trip to town before knowing they would be at home, I wrote them a note announcing that I would arrive at their house after lunch on National Day, but that if they had other plans, not to wait for me. On that day, as I was eating lunch with all my newspapers spread out on the table before me, Dr. Lin and his wife came to see me. She gave me some crabs and pastries and said, "Since it is too difficult for you to visit us, we had an early lunch so that we could come see you!" When they had sat down on my small benches, Dr. Yu noticed all my newspapers and exclaimed, "You work so hard!"

"I am not actually reading all of them," I replied. "I simply want to see what this Zhang character is lying about now." Dr. Yu asked if I was talking about Zhang Yongsheng. I said that I meant Zhang Chunqiao, one of the Gang of Four. Dr. Lin agreed that Zhang was always concocting wild stories, but when he left, he quietly told me by the door that I should be more careful of what I said. They did not allow me to see them off downstairs. That night I realized I should heed Dr. Lin's kind warning. If the authorities had heard my criticism of Zhang, I could have been imprisoned or even executed!

I sensed that the end of the troubled times was imminent. One day when Mo Xiquan picked up my prescriptions from Dr. Lin, he asked if I was in good spirits. Xiquan replied, "Yes! She is extremely cheerful!" At the beginning of October, people began to talk about the ruin of the Gang of Four, even though the official news had not yet been released. Everyone was excited about the promising future of our country since the years of disaster seemed to be over. For a long time, the Gang of Four had caused terrible destruction in China, but although we knew it would not be easy to get rid of their influence, we honest people were not going to allow the catastrophic years to be repeated. I continued to work at the library every day, but there were no longer any posters denouncing me.

Chapter 6

IN 1976 the diary of Lu Xun was published in a version more complete than earlier editions. As I was walking home one day, a colleague said, "You must have known Mr. Lu Xun personally, because he mentioned you in his diary! You should borrow a copy to read!" My job in the library did not assure me of being allowed to borrow books, but I somehow managed to acquire a copy of the diary for three days. Several days later, two women comrades from the Lu Xun Research Center in Beijing, comrades Ye Shusui and Zhao Shuying, came to persuade me to write an article in Lu's memory and to visit his museum in Beijing, which Ye Shusui had headed for many years. I sat down to write after they left, but each time I lifted my pen, my mind was filled with thousands of memories, and I did not know where to start. I had been determined to live by Lu Xun's famous words: "Fierce-browed, I calmly defy a thousand pointing fingers; head bowed, like a willing ox I serve the children." But I had been unworthy of his guidance and had become a rightist.

That year, old friends, colleagues, and relatives alike came to visit me. Some only stayed for a few days, some stayed for as long as two weeks. Ya'nan often had to sleep on a desk in my room. Among my visitors were Li Xin, Li Zhilian, Chen Chushu, and my niece and her husband. Dezhi's son Sang, his wife Man, and his son Ning also came to stay with me. When we wanted to talk about important matters, we always sent Ning to Ya'nan's room.

Sang then told me about his mother's death. She had been sick with diabetes, but the doctor had given her the wrong medicine. In her last days, her husband Mao Dun had spent his time running frantically up and down the stairs to tend to her.

Even though I had been "liberated," I was still under surveillance in the school. I no longer had to write reports on my activities, but every three or four days, the rebels would ask me if I had had any visitors. Most of the people who came to see me were studying or editing Lu Xun's works.

Many of my close friends had passed away. One day, some rebels from the textile industry where Zhang Qinqiu had worked came to question me about her. They told me that she had been optimistic and friendly with everyone and that no accusatory meetings had ever been held against her. One day, though, she saw a letter criticizing her and committed suicide by jumping out of her window. I told the rebels that I had nothing to say against her and nothing to hide from them. She had joined the revolution long before I did, and we had gone through the Yan'an rectification together in 1942. After Liberation, all our actions had occurred in full view of the party and the people. The rebels also asserted that my good friend Yang Zhihua had committed suicide, but, not believing them, I suspected they had killed her.

After many persistent invitations from the people at the Lu Xun Research Center, I decided to go to Beijing with Ya'nan on October 4, 1977. Since tickets for hard-class sleepers on the train were not easily obtained, Comrade Xu Wenyu, who worked in a railway station, arranged our transportation. He had suffered a great deal during the Cultural Revolution, because I, a notorious rightist, was among those who had sponsored him to join the party. In those days, he ran away whenever he saw me on the street to avoid talking to me, and I did not blame him. Since he had always worked tremendously hard, he survived the difficult times. One day when it was all over, he came to my house and was extremely emotional about our terrible experiences. Since then, our families had been close, and Ya'nan often went to his house to see his wife and daughters.

Chen Xuanzhao had always wanted me to come to her house in Beijing. After she retired from work, she fell and broke a thigh bone, so she could not walk very quickly. On the train to Beijing, I lay on my bed and read the *Chronology of the Life of Lu Xun*. Once when I was in the back of the train stretching my legs, a middle-aged man picked up my book and asked Ya'nan, "Is that woman your mother? What is your last name?" When Ya'nan replied that our last name was Chen, he said, "Oh! She must be Chen Xuezhao. I knew she was not a housewife." When I returned from my walk, he asked the reason for our trip. I explained that we were visiting relatives, which was partly true. He then informed me that he was the husband of a certain woman comrade and was going to Beijing on business. His wife, the party branch secretary who had ordered me to recruit the female student C. for the party, had labored in a village with me during the land reform and had told him that I never worked but always read French books until the wee hours of the morning. I smiled but said nothing when he told me about his wife. He then spoke about his adventures during the Cultural Revolution and about C., who still liked to flatter her superiors.

I had not traveled by this route for fifteen years. As the train passed Linping, Xieqiao, and Chang'an, I was reminded of when I had studied in Shanghai many years before. Entranced, I watched the houses, fields, and mulberry trees pass by and felt an emptiness in my heart. When we stopped in Shanghai, Ya'nan's husband and her little daughter were waiting for us. The child begged to come with us, so we had to promise her that we would return soon. The next evening we arrived in Beijing and were greeted by Xuanzhao's second son. We took a taxi to their house and ate dinner with the entire family. Since there was a shortage of many foods in Beijing, we brought quite a few provisions with us, such as rice and soybean oil. We had originally planned to stay at Xuanzhao's house for only two or three days before moving to the room that the research center had arranged for me, but when Xuanzhao heard this, she became angry. She exclaimed, "Do you not recognize me as your sister? I will not allow you to leave!" I explained that I had not wanted to cause her any trouble since she and her husband had many children to feed and look after, but we agreed to remain in the room she had provided for us on the third floor of her house. After the earthquake in Tangshan, many people in Beijing had erected small tents outside their houses in case of an emergency. One night, we heard a siren and went to sleep outside.

When Comrade Li Helin from the research center called on me one day, we were talking in the living room when comrades Sha Ting and Ding Ning, informed of my visit by Comrade Feng Mu, walked in. We all embraced and could not hold back our tears. For seven and a half years during the Cultural Revolution, Sha Ting had been imprisoned in Sichuan, where he lost his hearing. As a result, he could only communicate with pen and paper. Not allowed to visit him when he was in prison, his wife passed away before he was released. Ding Ning and her husband had been expelled from the Writers' Association and sent to toil in one of the worst villages in Hubei Province. She had just returned to Beijing, but he remained in the village. My two old friends gave me the address and telephone number of Zhou Yang and his wife Su Lingyang, and then departed at dusk.

The next morning, I attended a conference on Lu Xun, and people from many universities and research centers around the country came to ask me questions. I answered them to the best of my ability, but since I had lived in Zhejiang for so long and could not speak without a strong accent, the people could hardly understand me. Luckily, Ya'nan had also come to the meeting and acted as my interpreter. The conference ended at noon, and we spent our next three days visiting Lu Xun's former residence and the Lu Xun Museum.

I took out some time to visit Lu Xun's son Haiying and his wife but did not dare mention the loss of his mother, Xu Guangping. Her death had been indirectly caused by Zhang Chunqiao. Zhang had cut out all mention of himself from Lu Xun's diary, and when Xu expressed her displeasure, he ignored her. Following this incident, Xu had a heart attack and passed away shortly after returning from the hospital.

After I phoned Zhou Yang and Su Lingyang to announce our visit, Ya'nan and I took the subway to their house in the suburbs. It was as if we hadn't seen each other for centuries! I could not imagine how difficult their life must have been during the Cultural Revolution, and how they had managed to survive. While Zhou Yang was in prison for nine and one half years, Su Lingyang was assigned to labor in a frontier province, where she contracted heart disease and other illnesses that made her entire body swollen. Neither knew of the other's predicament or whereabouts. Ya'nan and I ate lunch with them but left shortly after, because they were still weak and needed to rest. I also called on Comrade Ge Baoquan, and we spoke at length of his deceased uncle, Ge Gongzhen. Ge Baoquan showed me a copy of *Ásya* by Ivan S. Turgenev, that I had translated and given to his uncle.

Dezhi's husband, Mao Dun, had already moved out of the Cultural Ministry's residence to a large house with two courtyards. I could sense Dezhi's presence throughout the house but did not mention her name to Mao Dun. He usually had lunch in his office, but that day we ate in the dining room. He ate very little before retiring to his room for a rest. I continued to chat with his son, Sang. When the boy told me that Dezhi's ashes remained in the house, I burst into tears and exclaimed, "I knew that she was still here!" He also told me that Qinqiu had not committed suicide but had been pushed out of her bathroom window at midnight. A month after Premier Zhou's death, Qinqiu's daughter had overdosed on sleeping pills after leaving a note that said that she could not bear Jiang Qing and Zhang Chunqiao. Sang was the only person who mourned her. The head of the cemetery decided he would not bury her, and her casket disappeared. When Ya'nan and I were ready to leave, Mao Dun's little granddaughter Dandan stood by the door and cried, "Please don't go, Granny, please don't!" I was deeply touched and promised her that I would be back soon. I could not see Mao Dun's children, Gang, who was in the Northeast, or Ning, who was not home.

During my stay in Beijing, comrades Ge Baoquan, Yan Rengeng of Beijing University, and Ye Yifen as well as my two nieces, came to see me. I refrained from calling on any of my friends, for I feared that when I returned to Hangzhou I would be criticized for visiting too many people. I never went out in the evening; instead, Xuanzhao and I sat around and talked all night. During the Cultural Revolution, her husband Wu Juenong had been beaten and paraded through the streets to show everyone that he was a capitalist-roader. The rebels had also taken all their savings from the bank. Before I left Beijing, I wrote to Deng Yingchao. Even though it had not revealed anything, I felt guilty that the rebels had confiscated her letter that I had forgotten to destroy.

We could see the famous author Lao She's house from Xuanzhao's window. Xuanzhao's family had a clear view whenever the rebels dragged Lao She out into the street for beating and kicking. One day, Xuanzhao saw some Red Guards escort him away from his house and he never returned. Two days later, it was reported that he had committed suicide. But the neighbors quietly told his widow that his body was found standing bolt upright in the very river bed he had written about in his 1951 play *Dragon Beard Ditch*. It was obvious that the Gang of Four had killed him and put him there because they thought that that particular play had praised Peng Zhen, one of their greatest enemies. In those times, the Gang of Four could murder someone in the blink of an eye!

Sang offered to help us obtain hard-class sleeper reservations for the trip home, so we went to his house to give Mao Dun the money for them and to give his little granddaughter some chocolate. The next day Ya'nan returned to pick up the tickets and brought back a box. In the box, underneath a bunch of Chinese wolfberries and haw chips, I found an envelope that contained two soft-class sleeper tickets, the best seats on the train, and all the money I had given Mao Dun. There was also a note from Sang asking my forgiveness for returning my money.

As soon as we arrived in Hangzhou, someone from the library came to ask if I had anything to report. I smiled and said that I had nothing to say. After several days, I began again to file newspapers, make cards for the catalog, and clean the rooms at the library. The New Year came and went. At the beginning of spring, Sang and his wife wrote to inform me that Gang had been admitted to the foreign languages department at Hangzhou University and would be in Hangzhou on March 8. By that time, the universities were no longer closed to all but peasants, workers, and soldiers, as they had been during the Cultural Revolution.

When Gang arrived, Ya'nan met her at the train station. After a quick lunch, Ya'nan borrowed a bicycle from a neighbor to carry Gang's luggage to the school and helped the girl settle into her dormitory. Three times a week, Gang would come to our house to eat dinner and listen to the radio, and Ya'nan would walk her halfway back to the school at 8:30. I thought Gang a lovely girl, and I was one of the few people who knew what hardships she had experienced. When she joined the party, she became a hardworking, sincere, and steady Communist. When others were talking, she would simply sit quietly and contemplate their words. Ya'nan and I were overjoyed to have her in the house, because we had been quite lonely in the previous twenty years.

Late in the spring of 1978, some of my articles were published for the first time in over twenty years, in *Shanghai Literature and Art* and *Zhejiang Literature and Art*. Fortunately, some of my rough drafts had survived the rebels' many searches of my house. Among the works that survived were my draft sequels to *To Be Working is Beautiful*, and *Spring Camellia*, and several fairy tales I had translated. I spent the entire hot summer revising the sequel to *To Be Working is Beautiful* and finally finished it. Before National Day, when the sweet osmanthus were about to bloom, I asked Ya'nan and Xiquan to accompany me to Lower Manjuelong, where I had worked many years before. The district had become very violent during the Cultural Revolution, with robberies and murders in broad daylight. This was my first of three visits to the village. There I finished the second volume of *Spring Camellia*. After 1957, especially in the ten years of the Cultural Revolution, my spirits had been so low that I had not noticed nature's beauty. Later, the white clouds over the green mountains reminded me of the days when I studied French in a girl's boarding school in the suburbs of Paris. I had lived on the second floor of my dormitory and had always thought of my motherland when I gazed out my window at the green mountains and the white clouds. I was amazed at how time had flown since I had written an essay over fifty years before entitled "The Mountains Are Green and the Clouds Are White" for the *Zhaohua xunkan* (Morning flowers fortnightly), whose editor-in-chief was Rou Shi.

Since the reputations of Wen Jize, Li Xin, and Li Zhilian as upstanding citizens had been reinstated, I awaited my turn. But nothing happened until February 27, 1979, when Ya'nan and I were invited to the premiere of the play *Song of the Wind* by Chen Baichen, whom I met at the entrance. He said that he had read my piece in the journal *Historical Materials on the New Literature** and from it had come to understand me for the first time. I smiled at him.

After the performance, Comrade Lin Danqiu kindly offered to drive my daughter and me to our house. As we passed through an intersection at a relatively high speed, a bus crossed through at a right angle without signaling or slowing down and hit us broadside. I was thrown to the floor of the car, and when Ya'nan helped me back up to the seat, I saw Lin lying unconscious with a gash on his head. Someone called for an ambulance to take the three of us to the hospital. After the doctor examined me, he concluded that I had damaged some nerves in my left leg and gave me several injections and some medicine. It was 2:00 A.M. before Ya'nan and I were allowed to go home. The next day, the pain in my legs and back were so sharp that I could not get out of bed, and Ya'nan could not stand up straight because of her backache. Therefore, Dr. Zhang Jiannan came to our house to attend to our discomforts.

One day Wei Bo, the deputy secretary general of the All-China Federation of Literary and Art Circles, brought three comrades to my house to discuss the restoration of my reputation. When he asked if I was in need of anything, I mentioned my hope that in view of my deteriorating health, Ya'nan's place of employment be close to my house, so she could help collect material for my writing and con-

---

*The February 1979 issue of this journal began the publication in installments of the first volume of the present memoir.

tinue typing my rough drafts. On the morning of March 4, while I was still bedridden from the car accident, several leaders from the provincial party Propaganda Department came to my house to show me a copy of their decision to rehabilitate me. It began: "Even though Chen Xuezhao made some inaccurate remarks during the rectification movement of 1957, she was not antiparty or antisocialist."

I handed the document back to them without signing my name, and said with a smile, " I would better understand my faults and be better able to reform myself if you would quote my misstatements."

One of the leaders returned my smile and replied, "Who can avoid misrepresenting the truth once in a while?" Everyone errs occasionally, I thought, but why was *my* fallibility written up in this official document that would be on file forever? Of course I did not express this opinion to them.

Shortly thereafter, a brief report on my rehabilitation and that of several others was published in the *Zhejiang Daily*, a local newspaper. But when the party had classified me as a rightist, there were long articles of criticism in all the major newspapers, including the *People's Daily*. I received quite a few letters advising me to request an equally comprehensive proclamation of my liberty, but I did not complain since I was not concerned with my personal gains or losses. If I had been interested in fame and fortune, I would not have chosen to work in Yan'an.

Although I was once again viewed as a loyal party member, I did not know where party meetings were held or to whom I should pay my dues. After an entire month, the leaders in the library still acted as if nothing had happened to me, so I decided to write the provincial leadership to ask where to submit my membership fee. After that, I attended two regular meetings at the library but had nothing to say to the people who had ostracized me during the Cultural Revolution. I sensed that some of them wished I would not attend the meetings. In May, notice of my rehabilitation finally filtered down to the university level. The party started to return the money that they had for years deducted from my salary, now amounting to nearly 700 yuan, and I paid my first party membership dues since 1957. But I was not recompensed in full until I retired from the school, since the leaders hedged about the issue each time I mentioned my finances. I was sure that they had just kept the money for themselves.

In late May, Gang transferred to the Beijing Languages Institute, because her grandfather missed her tremendously. Ya'nan and I were so accustomed to her visits to our house that often we still waited for her to arrive for dinner, especially on Saturday and Sunday nights. We had enjoyed her company a great deal and were lonesome when she returned to Beijing.

In 1979, I spent yet another summer in my tiny, sweltering room. Among the works of mine that were republished to celebrate the country's thirtieth anniversary were *To Be Working Is Beautiful, Spring Camellia*, and a collection of my stories, illustrated by Zhao Zongzao and with the title inscribed in the hand of Mao Dun. I was extremely appreciative of their help. But I was not very satisfied with my sequel to *To Be Working Is Beautiful*. I was still hedged in by inhibitions and conventions. It was only three years since the Gang of Four had been smashed. I hoped to revise it in the future.

Since I had no official title or political power, the provincial administration decided one day that my rooms should be given to someone else in the library. Comrade Guan Mingruo's attempt to persuade the administration otherwise was ineffective, so he sent me to look at another house. The previous owner, a hydraulic engineer, had been promoted and had moved to a larger house. The house had three extremely old and unclean rooms, a kitchen, and a bathroom, but knowing how difficult it had been for Guan to find even a run-down house, I accepted it.

Ya'nan, her husband, and her colleagues borrowed a truck to help me move. The building administration did nothing but collect the rent each month. It was said that assistance was only given to those with influence and connections, but I disagreed because Guan and others gave me a lot of help and expected nothing in return. I could not believe in such a general rule, as I was of the opinion that people's habits and morals governed their treatment of those less important than they.

Immediately after I settled in, I had to leave for the Fourth National Cultural Representatives Meeting in Beijing. At the request of the Provincial Federation of Literary and Art Circles, Ya'nan was allowed to accompany me. All the delegates from Zhejiang were in a special compartment of the train. We were greeted in Beijing by Li Xin, who appeared younger and healthier than when I had last seen him. I was able to chat with many colleagues whom I had not seen for a long time, because all the representatives from Zhejiang, Shanghai, Jiangsu, Jiangxi, Fujian, and Anhui stayed in the same hotel.

In Beijing, Zhou Yang and Zhang Bangying came to see me. During the Cultural Revolution, Zhang's lumbar vertebra had been broken on one of the occasions when the rebels had beaten him. Many of my other friends, such as Li Zhilian, Wen Jize, and Li Xin, also called on me. One day, Comrade Guo Mingqiu and her daughter Genggeng visited me, and we discussed the death of Lin Feng, Guo's husband, which had been caused by persecution at the hands of the Gang of Four. Mingqiu was still unusually weak from the ten years of adversity and had to lean back on the bed since she could not sit up straight. Genggeng, who bore a strong resemblance to her father, also had not yet thoroughly recovered from the physical hardships of the Cultural Revolution.

On the evening of October 29, we went to the opening

ceremony of a meeting in the Great Hall of the People. Hu Yaobang was the first speaker; he was followed by Deng Xiaoping, Zhou Yang, and Mao Dun. Several of the leaders who had over the years given me invaluable assistance and guidance were at the meeting, but I was most pleased to see Premier Zhou's widow Deng Yingchao, and comrades Peng Zhen, Kang Keqing, and Li Bozhao. I was exceedingly encouraged by the success of the conference. I wanted in my writing to sing the praises of the Four Modernizations, because I felt that all arts and cultural workers should unite under the excellent leadership of the party and work together for the future. The sight of these old friends and leaders was disheartening as well as encouraging, because even though the disastrous times were finally over, I knew I would never again see the many others who had passed away during those unfortunate years. We held memorial services for many of our beloved comrades, such as Feng Xuefeng, Zhou Libo, and Wu Xiaowu.

After calling on Chen Xuanzhao's family, I took a taxi to visit Yan Rengeng and Ye Yifen at Beijing University. The campus had changed a great deal during the Cultural Revolution; piles of garbage had appeared where trees had previously grown. Even though ten years of suffering had brought him heart disease, Yan Rengeng was quite happy because his party membership had just been reinstated. He was extremely proud of his only daughter, who loved to study and had recently been admitted to the university.

Since I knew my friends would say I was unreasonable if I did not do so, I finally resolved to visit Mr. and Mrs. Cai Wuji, the brother and sister-in-law of my old boyfriend Cai Boling. Wuji and I greeted each other as old friends, and he related some of his painful experiences from the Cultural Revolution. If Premier Zhou had not helped him, Wuji would not have survived. When I told him that all my French books had been confiscated, he asked me to give him a list of books I would like to read. About his brother, he said, "Boling is extremely depressed. He reads from morning until night, occasionally turning on the television. He no longer has an interest in maintaining his yard or in returning to Paris. Each time he is invited to an international conference to discuss physics or nuclear power, he recommends someone else for the job. He misses China exceedingly, and I wish he would come back to visit." Wuji became choked up, so I remained silent, but I did sympathize with Boling and felt a bit of guilt. Of course, I also thought of my other sweetheart, Mr. Ji Zhiren. In earlier days, I had repeatedly promised both of them that I wished to spend my life abroad rather than return to China. But in the end, I remained in China, and it was they who never returned since they knew that as intellectuals, they had no future here. I was unaware of Ji Zhiren's whereabouts until one of his cousins wrote to tell me that he was the director of a library in a university in the United States. But she had not heard from him at all during the Cultural Revolution, because the rebels had intercepted all their correspondence.

Following the cultural conference, I spent a week at Mao Dun's house while he was in the hospital. Before leaving Beijing, Ya'nan and I went to see him. Like a small child, he repeatedly remarked that he wanted to go home from the hospital. After I returned to Hangzhou, I received several French books from Cai Wuji and finally decided to send a short note and a copy of *To Be Working Is Beautiful* to Cai Boling in France. I then tried to concentrate on writing my memoirs.

In December, when Zhou Yang and Su Lingyang were in Hangzhou for a meeting, Lingyang stopped by to see me. When she remarked that my apartment was cramped and stuffy, I asked her if she had seen my previous lodging. She understood my point and said, "Oh, it is fine here. You even have a kitchen and a toilet!"

On January 2, 1980, the pain in my hip drastically

worsened. I suffered whether I walked, stood, sat, or lay down, and the acupuncture treatment that had previously been quite effective no longer abated the pain. An X-ray showed that I had hyperplasia in one of my vertebrae. It was so inflamed that it touched my sciatic nerve. On January 26, I finally checked into a hospital for government and party officials, the same one in which I had been treated for high blood pressure in 1955. But this stay in the hospital was completely different, because the service and the treatment were of much poorer quality. At first, I shared a room with four other patients, two of whom were highly contagious with influenza. After Guan Mingruo negotiated with the hospital staff, I was transferred to a small room that I shared with only one woman. Like me, she was used to sleeping in the day and reading at night. Since most patients did not really have physical illnesses, the hospital was more like a sanatorium. But I never rested well, because all the patients chatted and watched television until very late every night. After a comprehensive examination, a doctor found that I was developing diabetes, so he put me on a diet, which enfeebled me even more. Since Ya'nan had just moved to work in a new location, she could not visit me on a regular basis.

March 5, 1980, was the fortieth anniversary of the death of Cai Yuanpei, the father of Wuji and Boling. Wuji had expressed to me his hope that his brother would return from France for their father's memorial service, but on April 18, I heard startling news of Wuji's own death. Ever since, there was a rumor that Boling would someday return to China, but I was always extremely skeptical.

Even though I had worked in the same library since 1960 and therefore had had limited exposure, I was aware of the torture the rebels had inflicted upon people. While in the hospital, I realized the extent to which the Gang of Four had affected the nation, because as an intellectual with no political influence, I did not receive even adequate treatment. The doctors and nurses neglected me, once even giving me the wrong medication. Luckily I recognized their mistake and did not take the medicine. But when the doctor discovered what I had done, he exclaimed, "You are not ill and do not need medication!" After this episode, when my roommate, an official, was released from the hospital, the same doctor gave her a bag full of various kinds of medicine and a box whose contents I did not know. No longer able to tolerate this preferential treatment for others, I went out into the hall to complain to the doctor. When he urged me to accompany him to a room so we could talk alone, I firmly insisted that we remain in the hall so other people could hear our conversation. To absolve himself of any guilt, he finally said that he had only been teasing me when he had proclaimed that I was not sick, but I deemed his excuse quite unsatisfactory.

By April 13, a virus had spread to almost every patient in the hospital. For ten days, I suffered from a high fever and lost a great deal of weight. My friends among the patients recommended that I leave the hospital as soon as possible, so I returned to my house. Since my fever had not completely subsided, I called the doctor who had previously come to my house to treat me, and he relieved my fever, my rash, and the problems with my digestive system. After I had finally recovered, I could only write for half an hour at a time before taking a break to rest.

In 1978 and 1979, I was fortunate enough to be able to talk with Yu Ruomu, Chen Yun, and Wang Guangmei, the widow of former president Liu Shaoqi. All three had been models I had striven to emulate. Since I had discovered my many weaknesses through study, I resolved to refine my comprehension of the works of Marx, Lenin, and Mao. I knew I could withstand any sort of criticism, since women in China, especially those like me who had met with many difficulties, were often insulted because of their subordinate status. Some people denounced me for praising a female rightist in *To Be Working Is Beautiful,* but I maintained that everyone might have his or her own opinion. The woman was trying to continue enjoying her work amidst all the criticism, but the people put rightists on the level of criminals and wished they would commit suicide.

All these events reminded me of the venerable Ge Gongzhen, who had died under the Kuomintang's persecution forty-five years earlier. When he was criticized, he never got angry but simply said with a smile, "It is just one of those things." I decided to follow suit and continue writing and working for the rest of my life. I knew that under no circumstances would I surrender!

June 16, 1980
Hangzhou

Glossary of Chinese Characters

Note: First names and pseudonyms designated by a letter have been included in this glossary.

Ai Qing 艾青

Ai Siqi 艾思奇

Ba Jin 巴金

Bai Lang 白朗

baojia 保甲

Beiji 此枢

Beixin 此新

Bi Jilong 畢季龍

Bi Ye 碧野

Biaobei 裱褙

Cai Boling 蔡柏齡

Cai Wuji 蔡無忌

Cai Yuanpei 蔡元培

Cao Canru 曹參如

Cao Ming 草明

Cao Xiangqu 曹湘渠

Chang'an 長安

Chen Baichen 陳白塵

Chen Chushu 陳處舒

Chen Jiangong 陳建功

Chen Shumiao 陳樹淼

Chen Shutong 陳叔通

Chen Shuying 陳淑英

Chen Shuzhang 陳淑章

Chen Xuanzhao 陳宣昭

Chen Xuezhao 陳學昭

Chen Xuezhao yanjiu zhuanji 陳學昭研究專集

Chen Ya'nan 陳亞男

Chen Yi 陳毅

Chen Yun 陳雲

Cheng Yanqiu 程硯秋

Chenjiadai 陳家垗

Chun cha 春茶

Chung Kai-lai (Zhong Kailai) 鍾開萊

ci 詞

Cuncaoxin 寸草心

Dagongbao 大公報

Dai Li 戴笠

Deng Tuo 鄧拓

Deng Xiaoping 鄧小平

Deng Yingchao 鄧穎超

Ding Ling 丁玲

Ding Maoyuan 丁茂遠

Ding Ning 丁寧

Dongbei ribao 東北日報

Dongfang zazhi 東方雜誌

Donghai 東海

Fan Changjiang 范長江

Fang Lingru 方令儒

Faxiang 法相巷

Feng Mu 馮牧

Feng Xuefeng 馮雪峰

Fu Sinian 傅斯年

Fuchen zayi	浮沉雜憶
Gang	鋼
Gao-Rao	高崗．饒漱石
Ge Baoquan	戈寶權
Ge Gongzhen	戈公振
Ge Qin	葛琴
Genggeng	耿耿
Gongzuozhe shi meilide	工作著是美麗的
Guan Mingruo	管明若
Guo Lanying	郭蘭英
Guo Mingqiu	郭明秋
Guoxun	國訊
Haining	海寧
Haiying	海嬰
Hang (County)	杭
Hangzhou	杭州
Hanzi	函子
He Mu	何穆
Hepingli	和平里

Hu Nan	胡南
Hu Qiaomu	胡喬木
Hu Yaobang	胡耀邦
Hua Ti	華倜
Huairen	懷仁
Huang Yanpei	黃炎培
Huang Yuan	黃源
Huangdun	黃墩
Ji Baotang	元寶堂
Ji Zhiren	季志仁
Jiang Qing	江青
Jiaxing	嘉興
Jiefang ribao	解放日報
Jin Tao	金韜
Juanlü	倦旅
Kang Keqing	康克清
Kang Zhuo	康濯
Kong Dezhi	扎德沚
Kunqu	崑曲

Lao She   老舍

Lei Jia   雷加

Li Bozhao   李伯釗

Li Dequan   李德全

Li Fuchun   李富春

Li Helin   李何林

Li Lisan   李立三

Li Xin   黎辛

Li Zhilian   李之璉

Liang Zhaoming   梁昭明

Liangxi   梁溪

Lianxiu   連秀

Liao Zhigao 廖志高

Lin Biao   林彪

Lin Chenfu   林辰夫

Lin Danqiu   林淡秋

Lin Feng   林楓

Lin Liguo   林立果

Lin Mohan   林默涵

Lin Nengwu　　林能武

Lin Yutang　　林語堂

Linping　　臨平

Liu Baiyu　　劉白羽

Liu Shaoqi　　劉少奇

Liu Tianxiang　　劉天香

Liu Zhiming　　劉芝明

liulangzhe　　流浪者

liumang　　流氓

Liutong　　六通寺

Longjing　　龍井

Lu Dingyi　　陸定一

Lu Fei　　逯斐

Lu Xun (Zhou Shuren)　　魯迅（周樹人）

Luo Ruiqing　　羅瑞卿

Ma Feng　　馬烽

Ma Guangchen　　馬光辰

Ma Yinchu　　馬寅初

Mai Xin　　麥新

Man 曼

Manjuelong 滿覺隴

Manzou jiefangqu 漫走解放區

Mao Dun 茅盾

Meijiawu 梅家塢

Meng Jiao 孟郊

Meng Jimao 孟繼懋

Mo Xiquan 莫錫全

Mogan 莫干

Mu Ying 穆英

Nanfeng de meng 南風的夢

Nie Rongzhen 聶榮臻

Ning 寧

Ouyang Shan 歐陽山

Ouyang Yuqian 歐陽予倩

Peng Zhen 彭真

Qiu Jin 秋瑾

qu (district) 區

Qu Qiubai 瞿秋白

Rong 容

Rou Shi 柔石

Ru meng 如夢

Sang 桑

Sanlian 三聯

Sha Ting 沙汀

Shao Quanlin 邵荃麟

Shaoxing 紹興

Shen Congwen 沈從文

Shen Yanbing 沈雁冰

Shen Zemin 沈澤民

Shenbao "Ziyoutan" 申報 "自由談"

Shenghuo zhoukan 生活週刊

Shibao 時報

Shihuqiao 獅虎橋

Sibu 四部

Siyanjing 四眼井

Song Yunbin 宋雲彬

Su Buqing 蘇步青

Su Lingyang 蘇靈揚

Sun Fuxi 孫福熙

Sun Fuyuan 孫伏園

Tan Qilong 譚啓龍

Tan Zhenlin 譚震林

Tao Xingzhi 陶行知

Tian Jian 田間

Tianya guike 天涯歸客

Tonglu 桐廬

Tudi 土地

Waibaidu 外白渡

Wang Guangmei 王光美

Wang Qi 汪琦

Wang Yunru 王蘊如

Wei Bo 魏佰

Wen Jize 溫濟澤

Wenhuibao 文匯報

Wenxuan 文選

Wu Han 吳晗

Wu Juenong　　吳覺農

Wu Xiaowu　　吳小武

Xiao Jun　　蕭軍

Xiashi　　硤石

Xieqiao　　斜橋

Xin Hua　　新華

Xin nüxing　　新女性

Xin wenxue shiliao　　新文學史料

Xinyue　　新月

Xu Guangping　　許廣平

Xu Ruiyun　　徐瑞雲

Xu Wenyu　　徐文玉

Xu Zicai　　徐子才

Xucun　　許村

Yan Rengeng　　嚴仁賡

Yan Weibing　　嚴慰冰

Ya'nan　　亞男

Yan'an　　延安

Yan'an fangwen ji　　延安訪問記

Yang Hansheng 陽翰笙

Yang Shangkun 楊尚昆

Yang Shuo 楊朔

Yang Zhihua 楊之華

Yanguan 鹽官

Yangzhou 洋洲

Yao Wenyuan 姚文元

Ye Qun 葉群

Ye Shusui 葉淑穗

Ye Yifen 葉逸芬

Yequ 野渠

Yi Bali 憶巴黎

Yiqiao 義僑鄉

Yu Dafu 郁達夫

Yu Debao 俞德葆

Yu Ruomu 于若木

Yuying 育英

Zhang Bangying 張邦英

Zhang Chunqiao 張春橋

Zhang Jiannan   張劍南

Zhang Jingxin   張景新

Zhang Qian   張茜

Zhang Qinqiu   張琴秋

Zhang Tianyi   張天翼

Zhang Xi   張禧

Zhang Yun   章蘊

Zhang Zongxiang   張宗祥

Zhao Chaogou   趙超構

Zhao Shuying   趙淑英

Zhao Zongzao   趙宗藻

Zhaohua xunkan   朝花旬刊

Zhejiang   浙江

Zheng Boyong   鄭伯永

Zheng Xiaocang   鄭曉滄

Zheng Zhenduo   鄭振鐸

Zhou Jianren   周建人

Zhou Libo   周立波

Zhou Qiaofeng   周喬峰

Zhou Ren   周仁

Zhou Yang   周揚

Zhou Zuoren   周作人

Zhoushan   舟山

Zhu De   朱德

Zhufeng   珠鳳

Zou   鄒

Zuo   左

Index

Chen Xuezhao, now in her eighties, lives in the coastal city of Hangzhou in China.

Jeffrey C. Kinkley is the author of *The Odessey of Shen Congwen* (1987) and associate professor of Asian studies at St. John's University in New York.